The Bhopal Reader

The Bhopal Reader

Remembering Twenty Years of the World's Worst Industrial Disaster

Bridget Hanna
Ward Morehouse
Satinath Sarangi

Other India Press
Goa, India

The Apex Press
New York, USA

The Bhopal Reader

Edited by Bridget Hanna, Ward Morehouse and Satinath Sarangi

Published jointly by:

Other India Press
G-8, St Britto's Apartments, Feira Alta, Mapusa 403 507 Goa, India
Tel: 91-832-2263306; Tel./Fax: 91-832-2263305
email: admin@otherindiabookstore.com Website:
www.otherindiabookstore.com

The Apex Press
(an imprint of the Council on International and Public Affairs),
Suite 3C, 777 United Nations Plaza, New York, NY 10017, USA
Tel./Fax within the USA: 800-316-APEX (2739)
Tel/Fax outside the USA: 914/271-6500
E-mail: cipany@igc.org Website: www.cipa-apex.org

US Library of Congress Cataloguing-in-Publication Data

Hanna, Bridget.
The Bhopal Reader: Remembering twenty years of the world's worst industrial disaster / Bridget Hanna, Ward Morehouse, Satinath Sarangi
p. cm.

ISBN 1-891843-33-8 (USA : hardcover) – ISBN 1-891843-32-X (USA : softcover) – ISBN 81-85569-70-3 (India : softcover)

1.Bhopal Union Carbide Plant Disaster, Bhopal, India, 1984.

2. Pesticides

Industry – Accidents – India – Bhopal. I. Morehouse, Ward, 1929- II. Sarangi, Satinath. III. Title.

HD7269.C452I5244 2005
363.17'91– dc22 2004065963

ISBN 81-85569-70-3 (India-softcover)
ISBN 1-891843-32-X (USA-softcover)
ISBN 1-891843-33-8 (USA-hardcover)

All net proceeds from the sale of this book will be used to support the struggle for justice by the Bhopal survivors.

Printed by S.J. Patwardan at MUDRA, 383, Narayan, Pune 411 030, India.

To the Bhopal survivors who have never given up the struggle for justice

A Note on Permissions

Many of the selections included in *The Bhopal Reader* are in the public domain or their authors are quite willing to have them used provided attribution is given, as we have done throughout the book. For those selections not meeting these conditions, we have made a good faith effort to contact their authors or publishers to request permission to use their material, although accurate contact information was not always available. We apologize to any authors we were not able to reach and express our thanks to all who contributed to this attempt to refresh public memory of the world's worst industrial disaster.

Table of Contents

2005 Dow must show cause why it should not be asked to make Union Carbide face trial, this application to the court exemplifies the corporate veils and shifting assets that make multinationals so difficult to bring to trial.

106 Bhopal Chief Judicial Magistrate's (CJM), Order of January 6 (2005)
This order, requiring Dow Chemical Company, USA to state why it cannot present the absconder UCC in the ongoing criminal case for the '84 disaster, is one of the first legal orders to directly involve Dow in Carbide's liability.

PART 2: TWENTY MORE YEARS

SURVIVING

111 *Continuing crisis*

112 Voices from the Aftermath (1985-1990)
These brief, personal testimonials by survivors illustrate the continuing hardship of the aftermath of the Union Carbide disaster.

114 Bhopal Lives: Sajida Bano's Story and Harishankar Magician's (1996)
These two vignettes by Suketu Mehta deal with the difficulties of making a life after the gas. Sajida Bano is a gas widow who has to leave home after the disaster; Harishankar Magician has to adjust his trade to his condition after exposure.

116 The Goldman Foundation Environment Award (2004)
This speech was delivered by prize winners and Bhopal activists Rashida Bee and Champa Devi Shukla when they received their "green nobels." Bi and Shukla, both survivors, have been running a trade union of gas exposed women workers in an office stationary production unit for eighteen years.

118 Anger and Denial on the Streets of Bhopal (2004)
Suroopa Mukerjee looks into the conflicts within Bhopal, between the sick and struggling of Old Bhopal, different activist groups, the decision-makers and bureaucrats of New Bhopal, and the blame game that keeps them all apart.

SICKNESS

125 *Endless ills*
Epidemiological and Experimental Studies on the Effects of Methyl Isocyanate on the Course of Pregnancy (1987)
Daya R. Varma, one of the few scientists to study the effects of MIC on living systems, conducted research on its relationship to reproduction, and its effect on pregnancy through tests on mice.

PART 3: CHANGES

Evasion

Do We All Live in Bhopal?

PREFACE

As the struggle for justice in Bhopal has continued relentlessly, year after year, vital elements of this unique movement have disappeared from public memory. As Milan Kundera, the Czech author, reminds us, the struggle of people against power is the struggle of memory against forgetting. One of the aims of this book is to "refresh" that memory.

But an even more compelling aim of this volume is to provide the essential texts for an informed understanding of the disaster and its long aftermath. We hope that it will be particularly useful for students, researchers, journalists and others including those who are trying to bring a halt to the continuing disaster in Bhopal and recurrence of similar tragedies in other parts of the world.

A book with such a wide scope is bound to involve many helpers. Among those who did so with their hands, hearts and minds were Claude and Norma Alvares, Gary Cohen, David Dembo, Maude Dorr, Samira Desai, Tim Edwards, Tom Keenan, Chandana Mathur, Mary Ellen McCourt, Carolyn Toll Oppenheim, Judi Rizzi, Indra Sinha and Rukhsana Rushdi.

The organizations with which we are associated also lent a hand in ways too numerous and variegated to describe. They are the Bhopal Group for Information and Action, Council on International and Public Affairs, The Human Rights Project at Bard College, Program on Corporations, Law and Democracy, and Sambhavna Trust Clinic.

Given the limitations of space and time and the mismatch between our ambition and competence it is quite possible that there are small and large gaps in this collection. While we thank all who helped us in the project, none of them bears any responsibility for what appears and does not appear in the pages following. That responsibility is ours alone.

In the final analysis, we hope that the growing number of current and potential victims of industrial and environmental hazards will use the records of the disaster in Bhopal as foreknowledge of the script that is likely to unfold – and as a handbook for resistance and survival in the growing empire of toxic capital.

Bridget Hanna Ward Morehouse Satinath Sarangi
Bhopal, December 2004

Introduction

In what was arguably the worst industrial disaster in history, a methyl isocyanate (MIC) storage tank at a neglected Union Carbide pesticide factory in the town of Bhopal, India began to leak shortly after midnight on December 3, 1984. MIC is the active ingredient in the pesticide Sevin, which was widely touted as a "safe" pesticide. At the time of the accident, the staff in the MIC division had been reduced by half, and none of the plant's six safety systems was functioning. Some were awaiting repair, while others had been turned off in anticipation of the factory's sale.

The estimates of people who died on the spot, or running away from the cloud of gas that night range from 2,000 to 10,000 – of those exposed and surviving – 200,000 to 500,000. The legacy of the accident has been a horrifying catalogue of birth defects, secondary symptoms and neglect. Union Carbide, now a fully owned subsidiary of Dow Chemical Company, has implied, and at times stated, that the accident was a result of sabotage by an unidentified employee. Charged with culpable homicide and other serious offences by the Indian government, Union Carbide and its Chairman continue to abscond from the ongoing criminal case in Bhopal.

The gas disaster itself was a historic event. As the world's worst industrial disaster, it is a duly cited cautionary tale about the dangers of modern chemistry and the exigencies of modern capital. But Bhopal is at least two disasters – and the second has only just begun to make headlines. Bhopal's second disaster encompasses the conspiracies of law and government that led to the settlement with Union Carbide; the failure to bring to trial any of the negligent company's officials; the bureaucratic nightmare of the compensation courts; the systematic suppression of medical research and treatment of exposure even as the health crisis worsens; the knowing contamination of local drinking water from chemicals abandoned at the factory site; and the unending denial of truth and justice from corporate and government officials, both in India and in the United States (home to Carbide and Dow Chemical, who purchased Carbide in 1999).

Although many in this world are better protected than the Bhopalis, we are all vulnerable to industrial contamination. The voices from the Bhopal gas disaster – whether they be survivors demanding justice and clean water or company spokesmen defending the corporate right to run large-scale chemistry experiments unfettered by legal liability – are heralds of the dialogues that will dominate the next century. Because of our shared vulnerability, these are conversations that we must all begin to understand – because sooner or later, we all will have to join them.

Culminating in an unprecedented offer by the Indian government to join with survivors in litigation currently underway in New York, the anniversary year was full to bursting with major victories for the survivors. Just to name a few developments: the remainder of the 1989 settlement funds was finally cleared for disbursement to survivors; survivor activists won the prestigious Goldman Prize for the environment; Amnesty International issued a report on the unresolved nature of the Bhopal disaster; Innovest, a strategic investing firm, called Dow a poor financial investment based on continuing liability in Bhopal; and piping of clean water to replace contaminated wells in nearby communities finally began (though the activity has yet to provide adequate supplies). Also, a hoax disseminated by the BBC on the occasion of the twentieth anniversary – that Dow had decided to take full responsibility for Bhopal and its aftermath – gave the world a taste of what victory for Bhopal survivors might look like if the biggest chemical company in the world finally acted responsibly. At the beginning of January 2005 – in perhaps the most significant event of them all – a Bhopal judge ordered Dow Chemical Company to inform his court why it could not present the Union Carbide Corporation, now its subsidiary and an absconder, before the court. Perhaps this will be the step that finally penetrates the corporate veil.

Because corporations, government actors, and most of the world had written off Bhopal as a closed book, these victories so late in the game carry enormous significance. They show that just as heads of state are beginning to be brought to trial in international courts based on exhumations of mass graves in their territories, chemical disasters and miscarriages of justice like Bhopal cannot be buried and will not be forgotten. Unlike the carnage left by an earthquake or a tsunami, the Bhopal gas disaster left a health crisis that was acute in the days immediately after it happened, but that is arguably worse today, as a result of water contamination and genetic damage that affects the next generation. Yet disasters that happen on a corporation's watch are treated nearly as natural disasters, as though they are unfortunate but unpreventable. This may be beginning to change. If anything can change it, it will be the memory of Bhopal and its aftermath.

The Bhopal Reader brings together these texts about Bhopal for several reasons. First to refresh public memory of the event itself – and how it has been represented – by reprinting and annotating landmark writing from across the years. Second, to create a handbook for research and activism, a guide through the history, literature, and politics of the gas disaster. And third, to contextualize and make legible the complex dynamics of the Bhopal story by drawing connections between, rather than reinterpreting, the many stories and statements that have brought Bhopal up to 2005.

However, we do not pretend that this is a comprehensive effort. Many important texts have been omitted for reasons of space, and certainly many other have been missed. Part of the continuing effort that this represents is one of creating an archive that will be a resource for future generations. For this reason, we encourage readers to contact the editors with any other texts or materials that should have a place in this ongoing archive.

That this anthology covers twenty years is not an accident of round numbers; rather it is an imperative brought on by what appears to be a change in the winds for the fate of Bhopal survivors (and hopefully for similarly affected communities in other countries) giving credence to our theory that the struggles following the accident are as unique and powerful as the catastrophe itself.

Because of the magnitude of the Bhopal disaster, the relative inaccessibility of the judicial processes to survivors (in either India or the United States), and the relative disinterest shown over the years by the medical, scientific and academic communities about the consequences of the event, the discourse about Bhopal has had a continued dynamism and complexity – fed by the continued legal struggles, the never agreed upon facts of the accident, and the medical crisis on the ground. Those who work for justice and health in Bhopal are struggling to heal not only the affected lives and communities, but also the systemic flaws that allowed the disaster to occur and that cause justice, in the eyes of the survivors, to remain a pending dream. *The Bhopal Reader* maps that fascinating and desperately important terrain.

Guided by the annotated table of contents, this book is meant to be both a reference text for industrial accidents and corporate crime and a handbook for prevention and activism. The *Reader* begins with a timeline – to help ground the complicated story that is told by the texts that follow. The anthology itself is divided into three parts, and is followed by a resources section. The first part – Disaster – looks at the lead up to the ghastly event; reports the experiences of survivors and explanations of why it took place; compensation (and the lack of it) and, finally, the legal issues surrounding the 1984 tragedy. The second part – Twenty More Years – is the long and painful story of the continuing aftermath of Bhopal. Finally, the last section – Changes – looks at the way Bhopal has become a global issue: both at how the world has come to Bhopal, and how Bhopal continues to change the world. Sections are divided into chapters: each chapter begins with a brief introduction from the editors followed by illustrative excerpts from both primary and secondary sources.

THE BHOPAL TIMELINE

1969 Union Carbide India Limited's (UCIL) plant at Bhopal designed by its holding company Union Carbide Corporation (UCC), U.S.A. (which held 50.9% of UCIL's equity), was built in 1969 as a formulation factory for UCC's Sevin brand of pesticides, produced by reacting Methyl Isocyanate (MIC) and alpha naphthol. Sevin kills pests by paralysing their nervous systems. At this time MIC was imported from the US in steel containers. Plant set up on land taken on long-term lease from State of Madhya Pradesh.

1975 UCC decided to "integrate backwards" and manufacture ingredients of Sevin at the Bhopal plant of UCIL. Although the then zonal regulations prohibited locating polluting activity in the vicinity of 2 kms from railway stations, UCC was able to persuade the authorities to grant it the necessary clearances.

1978 The alpha naphthol manufacturing unit was set up. A year later, the MIC unit was also erected at UCIL's plant in Bhopal.

December 25, 1981 Leak of phosgene gas at the UCIL plant killed one worker.

January 9, 1982 25 workers were hospitalised as a result of another leak at the UCIL plant. Workers' protests went unheeded.

1982 Bhopal journalist Rajkumar Keswani wrote a series of articles in local press about the dangers posed by the UCIL plant.

March 1982 A leak from one of the solar evaporation ponds took place. In April 1982 a UCIL document addressed to UCC noted that the continued leakage was causing great concern.

May 1982 UCC sent its US experts to the UCIL plant to conduct an audit. The team noticed leaking valves and a total of 61 hazards, 30 of which were major and 11 of which were in the MIC/phosgene units.

September 1982 UCIL de-linked the alarm from the siren warning system so that only its employees would be alerted over minor leaks without "unnecessarily" causing "undue panic" among neighbourhood residents.

October 5, 1982 Another leak from the plant resulted in hospitalisation of hundreds of nearby residents.

March 4, 1983 Bhopal lawyer Shahnawaz Khan served a legal notice on UCIL stating that the plant posed a serious risk to health and safety of workers and residents nearby.

April 29, 1983 In a written reply, UCIL's Works Manager denied the allegations as baseless.

Between 1983 & 1984 The safety manuals were rewritten to permit the switching off of the refrigeration unit and the shutting down of the vent gas scrubber when the plant was not in operation.

The staffing at the MIC unit was reduced from 12 to 6.

Thus, at the time of the disaster on the night of December 2/3, 1984,

- The refrigeration unit installed to cool MIC and prevent chemical reactions had been shut for 3 months

- The vent gas scrubber had been shut off for maintenance
- The flare tower had been shut off
- There were no effective alarm systems in place
- The water sprayers were incapable of reaching the flare towers
- The temperature and pressure gauges were malfunctioning
- Tank number 610 for storing MIC was filled above recommended capacity

The stand by tank for use in case of excess was already having MIC.

November 29, 1984 UCC had by this time decided to dismantle the plant and ship it to Indonesia or Brazil. The feasibility report for this was completed on November 29, three days before the disaster.

December 2, 1984 At 8.30 p.m. the workers, under instructions from their supervisors, began a water-washing exercise to clear the pipes choked with solid wastes. The water entered the MIC tank past leaking valves and set off an exothermic "runaway reaction" causing the concrete casing of tank 610 to split and sending the contents leaking into the air.

December 3, 1984 Because of no warning given to residents or about precautions they should take, many of them ran on to the streets to meet a certain death. A suo-motu FIR was recorded by the Station House Officer at Hanumanganj police station on 3.12.84 against UCC, UCIL and its executives and employees under s.304 A IPC. The record indicates the grim statistics:

- 3828 died on the day of the disaster (the unofficial toll is feared to be much higher – by 2003 over 15,000 death claims have been processed)
- Over 30,000 injured on the fateful day (a figure that now stands at 5.5 lakhs)
- 2544 animals killed

December 3, 1984 Five junior employees of UCIL were arrested.

December 6, 1984 The case was handed over to the CBI.

December 7, 1984 Warren Anderson (A1), Keshub Mahindra (A2) and V. P. Gokhale (A3) were arrested and released on bail the same day. A1 was escorted out to Delhi on the Chief Minister's special plane.

Nearly 145 claims were filed on behalf of victims in various US courts. These were consolidated and placed before the Southern District Court, New York presided over by Judge John Keenan.

March 29, 1985 Parliament enacted the Bhopal Gas Leak Disaster (Processing of Claims) Act 1985 whereby Union of India would be the sole plaintiff in a suit against the UCC and other defendants for compensation arising out of the disaster.

April 8, 1985 Union of India filed a complaint on behalf of all victims in Keenan's court.

October 29, 1985 UCIL which was still in control of the plant (except the MIC unit which had been sealed by the CBI) wrote to UCC that clean up was going on but "some material remains in the tank bottom."

December 1985 CSIR submitted a detailed report squarely implicating the UCC for faulty design of the plant as well as its reckless disregard of operational safety.

May 12, 1986 Accepting the *forum non conveniens* defence, Judge Keenan dismissed the Indian Government's claim, conditional upon UCC submitting to the jurisdiction of the Indian courts.

September 5, 1986 Union of India filed a suit against UCC in the Bhopal District Court.

December 1, 1987 CBI filed a charge sheet in the court of the Chief Judicial Magistrate, Bhopal charging accused for offences under s. 304 Part II IPC and other offences.

December 17, 1987 An interim compensation of Rs.350 crores was ordered by Judge Deo, District Judge, Bhopal.

April 4, 1988 This was challenged before the High Court at Jabalpur. By judgment dated April 4, 1988 the High Court reduced the interim compensation to Rs.250 crores. UCC challenged this further before the Indian Supreme Court.

February 14/15, 1989 The Supreme Court approved a settlement arrived at in the appeal by UCC whereby $470 million was to be paid by it and UCIL to the Union of India in full and final settlement of all claims. Criminal proceedings would stand quashed.

May 4, 1989 Following widespread protests over the manner of arriving at the settlement and the quashing of criminal proceedings, the Supreme Court agreed to review the settlement.

December 22, 1989 The Supreme Court upheld the validity of the Claims Act applying the doctrine of *parens patriae.*

June 1989 Meanwhile UCC in June 1989 finalised a "Site Rehabilitation Project – Bhopal Plant" for decontamination of the plant site which contained huge quantities of Sevin and naphthol tarry residues and solid wastes dumped in the solar evaporation ponds. Since no Indian organisation had the expertise, it was decided to get NEERI to undertake the task under the supervision of Arthur D. Little & Co. appointed by UCC.

1990 NEERI submitted its first report in 1990 stating that there was no contamination of the groundwater in and around the plant site. Subsequent documentation reveals that UCC itself doubted NEERI's conclusions since their internal notes revealed that majority of liquid samples collected from the area "contained naphthol or sevin in quantities far more than permitted by ISI for inland disposal."

October 3, 1991 The Supreme Court declined to reopen the settlement justifying it under Article 142 of the Constitution. However, the criminal proceedings were directed to be revived. The court expressed a hope that UCC would contribute Rs.50 crore to setting up a hospital at Bhopal for the victims.

February 1, 1992 The CJM Bhopal declared A1 Warren Anderson A10 UCC and A11 UCC (Eastern, Hong Kong) as proclaimed offenders. The CJM directed that if parties do not appear on March 27, 1992 he will order attachment of UCC's shares in UCIL under s.82 Cr.PC.

March 27, 1992 A1, A10 and A11 fail to appear before the CJM but attachment of shares was put off at UCIL's request.

April 15, 1992 UCC announced creation of the Bhopal Hospital Trust in London with Sir Ian Percival as Sole Trustee and endowed its entire shareholding in UCIL to the Trust clearly to defeat the attachment.

April 30, 1992 CJM Bhopal refused to recognise the creation of the Trust and endowment of UCIL shares and proceeded to attach those shares.

June 22, 1992 Trial of the Indian accused was separated and committed to Sessions Court.

April 8, 1993 Charges framed by the Sessions Court, Bhopal against Indian accused for offences under s.304 Part II IPC.

December 10, 1993 Ian Percival approached the Union of India with an "offer" to sell the attached shares of UCIL to raise money for the Bhopal Hospital to be built by UCC. Union of India filed an application in the Supreme Court for enforcement of UCC's obligation to build the expert medical facility. At the first hearing of the application, Ian Percival was present and heard. The Supreme Court asked Union of India to consider the Sole Trustee's suggestion which was "eminently reasonable, worthy of consideration."

February 14, 1994 The Supreme Court modified the CJM's attachment order and permitted the attached shares to be sold.

September 1994 UCC's shares in UCIL sold to McLeod Russel Ltd. for Rs.170 crores. UCIL renamed as Eveready Industries India Limited (EIIL). After release of around Rs.125 crores (inclusive of dividends) to the BHT, the balance sale proceeds to the tune of about Rs.183 crores remained under attachment.

April 3, 1996 Supreme Court directed a sum of Rs.187 crores from the attached monies to be further released to BHT for the construction of the hospital.

September 13, 1996 Meanwhile Indian accused failed in their challenge to the order framing charges before the High Court at Jabalpur. They then approached the Supreme Court by way of Special Leave Petitions. By judgment dated September 13, 1996, the Supreme Court diluted the charges against the Indian accused from s.304 Part II IPC to s.304 A IPC. The trial is still pending before the CJM, Bhopal.

October 1997 EIIL retained NEERI and Arthur D. Little to conduct further decontamination studies. NEERI submitted its second report, again maintaining that there was no contamination of groundwater and soil around the plant site. But Arthur D. Little did not rule out such contamination.

September 1998 State of M.P. took control of the land and put up notices in nearby residential areas warning against drinking water.

November 1999 Greenpeace, an international environmental NGO, came out with an independent report of test of soil and water samples collected in areas around the plant site and confirmed extensive contamination. Sevin was seen to have leaked from the ruptured tank, and water supplies were found contaminated with "heavy metals and persistent organic contaminants."

November 15, 1999 Fresh class action litigation filed in the court of the Southern District New York by Sajida Bano, Haseena Bi and five other victims directly affected by the contamination and five Bhopal victims groups claiming damages under 15 counts. Counts 9 to 15 related to common law environmental claims.

August 28, 2000 Judge Keenan dismissed the class action claim on the ground that the 1989 settlement covered all future claims.

February 5, 2001 The US Federal Trade Commission approved the merger of UCC with Dow Chemical Company (Dow).

November 15, 2001 The Second Circuit Court of Appeals affirmed in part but remanded claims on counts 9 to 15 to Judge Keenan.

April 2002 In discovery proceedings before Judge Keenan, UCC submitted over 4000 documents.

March 2003 Judge Keenan dismissed the suit of Hasina Bi again – this time on grounds of limitation.

March 2004 The Court of Appeal affirmed in part, but asked Judge Keenan on remand to consider claims of Bi arising out of damage to property and the issue of decontamination by UCC of the site if the Union of India and State of M.P. had no objection.

June 28, 2004 After victims went on a hunger strike in Delhi, the Union of India submitted a memo before Judge Keenan stating it has no objection to decontamination being undertaken by UCC at UCC's cost.

July 19, 2004 In a representative application by 36 victims (Abdul Samad Khan and others), the Supreme Court directed disbursement of balance compensation.

September 17, 2004 In another writ petition by the Bhopal groups for medical relief and rehabilitation, the Supreme Court finalised the terms of reference of two committees appointed by it.

October 26, 2004 In the Abdul Samad Khan application, the Supreme Court directed the disbursement of balance compensation to commence from November 15, 2004 and to conclude by April 1, 2005. The court accepted the Action Plan prepared by the Welfare Commissioner.

On the 20th anniversary of the disaster, the BBC broadcast a story that the Dow Corporation had decided to accept responsibility for the Bhopal disaster, push for the extradition of Warren Anderson, and had set aside $12 billon to compensate victims and rehabilitate the factory site. An hour later, the story that has been broadcasted internationally, is revealed to be a hoax by the activist duo, the Yes-men.

January 6, 2004 The Chief Judicial Magistrate in Bhopal, following a petition by activist groups, ordered that, since the absconding Union Carbide Corporation is a fully owned subsidiary of Dow Chemical Company, Midland, Michigan, USA, this company be made a party in the criminal case. A notice was issued to Dow USA, and the case was fixed for February 15, 2005.

Source: S. Muralidhar, from *Seminar 544: Elusive Justice*, December 2004.

PART 1: DISASTER

1

The Night of the Gas

December 3, 1984

THE LEAKAGE OF some of the most lethal chemicals known to humankind began without the knowledge of the mass of people they were to kill, or maim for life. Most were sleeping and no sirens were sounded to warn them. As people woke up, gasping, choking, coughing, nobody knew which way to run to escape the lethal effects. Nobody knew that it was better if you didn't run from the burning fumes, that the harder you breathed, the worse it was for your body. Nobody knew that a wet cloth over the eyes and nose and mouth offered a protection that might save your life. Nobody knew what was leaking from the Union Carbide pesticide plant; most did not even know what was produced there. There was no emergency plan. The leak of 40 tons of methyl isocyanate (MIC), began just after midnight. At 2 a.m. the Union Carbide spokesman in Bhopal was still denying to the local police that there was anything wrong at all. In short, there was absolutely zero concern by the Union Carbide Corporation for the lives of the people of Bhopal: not before the accident, not during the accident, and certainly not after the accident. None.

The survivors of the gas disaster lost their family and friends, their health and thus their means to livelihood, all in one unforgettably and at the same time unspeakably mysterious and terrifying night. They are, and will remain, haunted by it. When the survivors emphasize that there should be "no more Bhopals" anywhere, they mean that never again should anyone have to live through the fear and misery of *that night* again. Any understanding of the disaster begins and ends with the stories of the women, men and children who survived *that night.*

Ramesh's story (1987)

A student and apprentice at a tailor's shop, Ramesh was 12 years old in December 1984 and living in Jai Prakash Nagar.

The day before the gas leaked was a Sunday. My friends and I were relaxing in the evening and then we watched a movie on the television. I must have gone to sleep around 9 p.m. It was cold and I had covered myself with a rug. Sometime in the middle of the night I heard a lot of noise coming from outside. People were shouting "Get up!" "Run, run!" "Gas has leaked!" My elder brother Jawahar got up and said, "Everyone is running away, we too must run!" I opened my eyes and saw that the room was full of white smoke. The moment I removed the rug from my face my eyes started stinging as if someone was burning a lot of dried chilies and every breath was burning my insides. I was scared of opening my eyes. The gas was getting in through my mouth. Through my nose. We got ready to run. All six of us, my brothers and sister came out together, with my sister carrying my youngest brother Rajesh in her arms. My father refused to leave and my mother stayed with him. So we left them in the house and ran towards the cremation ground.

After a while my sister who was carrying my little brother got separated. The gas was thick and we couldn't see where they had gone. The four of us left, held on to each other's hand and walked on. While running Mahesh and I fell into a ditch full of dirty water. As we reached the main road we could see a lot of people lying around. We did not know whether they were dead or unconscious. One fellow, Gupta, was sleeping with a rug over him. Mahesh crawled under the rug with him and that fellow made place for the rest of us. But after a while as we could not breathe it got very uncomfortable under the rug. So we got up and started walking towards the bus stand. Near the wine shop we came across some bathrooms which were closed. My brother kicked the door down and all of us got inside. I covered my younger brother Mahesh and Suresh with my coat and put them to sleep. My elder brother kept sitting outside looking out for our parents. I asked him to come in but he was worried about them and would not come in. He started vomiting outside. Some people in the neighborhood gave him water. We were not talking to each other just sitting and worrying about our parents. I did not know that the gas could kill people but while running away I had seen a little child crying beside his mother who was lying beside the road. So all kinds of dreadful thoughts crossed my mind.

Early in the morning we set out for home. My eyes were swollen and my chest was aching. On our way back we saw a lot of dead cattle lying around and a lot of people too. My brother could not walk. So Mahesh and I held both his hands and pulled him along. Nearer home I saw my friend Santosh's grandfather lying dead. Balmukund, our neighbor was dying and they rushed him to the hospital. My uncles who also live in a *bustee* in Bhopal had heard about the gas leak and came looking for us. They were worried about what they would find. But they found us all alive. As I was going with my uncle to buy jaggery, on the way to the shop I saw a lot of dead bodies of men, women and children lying in front of the Union Carbide factory gate. I was also trying to find our cow and found her in a street coughing. My dog lay dead. Two of my friends Santosh and Rajesh also had died. Then my uncle took all of us to their home which is fifteen kilometers away. We sat under a tree and all the people in the locality came over to look at us. Then people from amongst them arranged to have us taken to the hospital.

Source: From "We Will Never Forget" BGIA, December 1987

Bano Bi's story (1990)

Bano Bi was a 35-year-old housewife living in Chhawani, Mangalwara the night of the gas disaster.

The night the gas leaked, I was sewing clothes sitting next to the door. It was around midnight. The children's father had just returned from a poetry concert. He came in and asked me, "What are you burning that makes me choke?" And then it became quite unbearable. The children sleeping inside began to cough. I spread a mat outside and made the children sit on it. Outside we started coughing even more violently and became breathless. Then our landlord and my husband went out to see what was happening. They found out that some gas had leaked. Outside there were people shouting, "Run, run, run for your lives!"

We left our door open and began to run. We reached the Bharat Talkies crossing where my husband jumped into a truck full of people going to Raisen and I jumped into one going towards Obaidullahganj. It was early morning when we reached Obaidullahganj. The calls for the morning prayers were on. As we got down, there were people asking us to get medicines put on our eyes and to get injections. Some people came and said they made tea for us and we could have tea and need not pay any money.

Meanwhile, some doctors came there. They said the people who are seriously ill had to be taken to the hospital. Two doctors came to me and said that I had to be taken to the hospital. I told my children to come with me to the hospital and bade them to stay at the hospital gate till I came out of the hospital. I was kept inside for a long time and the children were getting worried. Then Bhairon Singh, a Hindu who used to work with my husband, spotted the children. He too had run away with his family and had come to the hospital for treatment. The children told him that I was in the hospital since morning and described to him the kind of clothes I was wearing.

Bhairon Singh went in to the hospital and found me among the piles of the dead. He then put me on a bench and ran around to get me oxygen. The doctors would put the oxygen mask on me for two minutes and then pass it on to someone else who was in as much agony as I was. The oxygen made me feel a little better. The children were crying for their father so Bhairon told them that he was admitted to a hospital in Raisen. When I was being brought back to Bhopal on a truck, we heard people saying that the gas tank has burst again. So we came back and went beyond Obaidullahganj to Budhni, where I was in the hospital for three days.

I did not have even a five paisa coin on me. Bhairon Singh spent his money on our food. He even hired a taxi to take me back to Bhopal to my brother's place. My husband also had come back by then. He was in a terrible condition. His body would get stiff and he had difficulty in breathing. At times, we would give up hopes of his survival. My brother took him to a hospital. I said that I would stay at the hospital to look after my husband. I still had a bandage over my eyes. When the doctors at the hospital saw me, they said: "Why don't you get admitted yourself, you are in such a bad state." I told them that I was all right. I was so absorbed with the sufferings of my children and my husband that I wasn't aware of my own condition. But the doctors got me admitted and since there were no empty beds, I shared the same bed with my husband in the hospital. We were in that hospital for one and a half months.

After coming back from the hospital, my husband was in such a state that he would rarely stay at home for more than two days. He used to be in the Jawaharlal Nehru Hospital most of the time. Apart from all the medicines that he used to take at the hospital, he got medicines like Deriphylline and Decadron from the store. He remained in that condition after the gas disaster. I used to take him to the hospital and when I went for the Sangathan meetings, the children took him to the hospital. He was later admitted to the MIC ward and he never came back from there. He died in the MIC ward.

My husband used to carry sacks of grain at the warehouse. He used to load and unload railway wagons. After the gas, he could not do any work. Sometimes, his friends used to take him with them and he used to just sit there. His friends gave him five to ten rupees and we survived on that.

We were in a helpless situation. I had no job and the children were too young to work. We survived on help from our neighbors and other people in the community. My husband had severe breathing problems and he used to get into bouts of coughing. When he became weak, he had fever all the time. He was always treated for gas related problems. He was never treated for tuberculosis. And yet, in his post-mortem report, they mentioned that he died due to tuberculosis. He was medically examined for compensation but they never told us in which category he was put. And now they tell me that his death was not due to gas exposure, that I can not get the relief of Rs.10,000 which is given to the relatives of the dead.

I have pain in my chest and I go breathless when I walk. The doctors told me that I need to be operated on for ulcers in my stomach. They told me it would cost Rs.10,000. I do not have so much money. All the jewelry that I had has been sold. I have not paid the landlord for the last six years and he harasses me. How can I go for the operation? Also, I am afraid that if I die during the operation, there would be no one to look after my children.

I believe that even if we have to starve, we must get the guilty officials of Union Carbide punished. They have killed someone's brother, someone's husband, someone's mother, someone's sister – how many tears can Union Carbide wipe? We will get Union Carbide punished. Till my last breath, I will not leave them.

Source: From "Voices from Bhopal" BGIA, December 3, 1990

Bhopal lives: The night of the gas (1996)

Journalist Suketu Mehta captures the human dimension of the tragedy in this, the first of a series of vignettes about the victims of Union Carbide's negligence.

Bhopalis have very personal relationships with "the gas." Accounts of that night – again, when in Bhopal someone says "that night," they mean the night of 2/3 December 1984 – describe how the gas was going toward Jahangirabad or Hamidia Road; how it hovered a few feet above the ground at some places or how it hugged the wet farm earth in others; how it killed buffaloes and pigs but spared chickens and mosquitoes; how it made all the leaves of a peepul tree turn black and how it had a particular hunger for the tulsi plant; how it would travel down one side of the road but not the other, like rain falling a few feet from you while you're standing in the sunshine. People know the gas like a member of their family – they know its smell, its color, its favorite foods, its predilections. One thing everybody remembers is the smell of chilies burning. Chilies are normally burned to ward off the evil eye, when, for example, a child is sick. People woke up and thought: it must be a powerful evil eye that's being driven away, the stink is so strong.

As people ran with their families, they saw their children falling beside them, and often had to choose which ones they would carry on their shoulders and save. This image comes up again and again in the dreams of survivors: in the stampede, the sight of a hundred people walking over the body of their child.

Iftekhar Begum went out on the morning after the gas to help bury the Muslim dead. There were so many that she could not see the ground – she had to stand on the corpses to wash them. As she stood on the bodies, she noticed that many of the dead women had flowers in their hair. The gas had come on a Sunday, a night when people had dressed up to go out to a film or to someone's house for dinner. The women had, as is common all over India, braided their hair with jasmine or mogra – small, fragrant flowers.

When Iftekhar Begum came back from the graveyard, all her fingertips were bleeding, she had sewn so many shrouds.

Source: *Village Voice*, December 3 and December 10, 1996.

2

Predictions

Could the gas disaster have been prevented?

THE BHOPAL GAS DISASTER was not a simple accident. It was caused by negligence; firstly in the construction of a dangerous facility using "unproven technology", and secondly through poor and irresponsible maintenance of that facility. Regardless of the exact circumstances of the disaster itself, it was clear, well before the event, that a catastrophe was imminent. Not only was the danger known, but the officials in charge of the plant had been explicitly and repeatedly warned. Few people knew much about the factory. But those who did – the workers at the factory and a few citizens and journalists – saw clearly what could happen. And they didn't keep quiet.

Although the factory had inherent design flaws, the early eighties had witnessed an additional decline in the safety culture at the Carbide plant. Seasoned engineers left because of disagreements over safety issues. Workers, concerned about the death of a fellow employee, Mohammed Ashraf, wrote to the company asking that it review safety measures and damaged equipment, but were told that there were no safety problems. Citizens voiced their worries – advocate Shahanawaz Khan for example wrote to Carbide in 1983, outlining nearly every danger that would come to characterize the accident. And consistently over the preceding years, Bhopali journalist, Rajkumar Keswani, published his concerns in the local papers, warning the people of Bhopal that the likely release from the factory would amount to a moving gas chamber. As late as 16 June 1984, less than six months before disaster struck, he published an exhaustive report on the time bomb at the Union Carbide plant in *Jansatta*, a Hindi-language national daily. Included below is his article written just after the disaster, in which he traces the story of how this inevitability developed.

But Keswani, who won a prestigious award for his prescient journalism, commented years later in an interview that he was rewarded for "journalism's most spectacular failure. Had I done my job and prevented the disaster, no one would know my name." In spite of their failure, these efforts at preventing catastrophe are now part of a crucial documentary case, and only reiterate the fundamental importance of learning and speaking about industrial dangers.

The Right to Know movement was the primary response internationally to the Bhopal disaster. But what Bhopal and Keswani's experience demonstrate most clearly, is that knowledge is not sufficient. Beyond the right to know, there must be the right, indeed the imperative, to act to prevent disaster. Sometimes, as in this case, paper is not enough.

It is also possible that nothing would have been enough to prevent the gas leak. As documents subpoenaed from Carbide during the discovery phase of a later litigation reveal, the factory in India was built with unproven technology, and was not required to adhere to the same safety standards as its "sister" plant in Institute, Virginia, USA. Additionally, as nearly everyone familiar with the substance agrees, storing the excessive amounts of methyl isocyanate at the Bhopal plant for long periods of time was like lighting the fuse on a forty ton dirty bomb.

How could I know what poisons were stored in there? (1990)

Ramkishan, 40, was a worker at the Carbide factory, and lived close to it.

I used to work at the formulation plant in Union Carbide's factory. I had been working there ever since the sixth month of the year 1973. When I joined, I used to work as a casual worker. For six years, I worked as a casual worker. They made me a permanent worker in the third month of 1980. I was working in the formulation plant on the night the gas leaked. The tank of MIC which leaked was only 400 feet from the formulation plant. When the gas started leaking, some people cried out, "Run, run!" and we left our work and ran towards the west...

I was just a worker there, how could I know what poisons were stored in there? I was never told that there were such dangerous chemicals inside the factory. If I knew, I would not have worked in that factory. The plant used to smell awfully at times but we were just workers, how could we know? When we worked there, our eyes used to hurt and our skin itched but whoever knew that such a disaster could happen?

Source: From "Voices from Bhopal" BGIA, December 3, 1990

Report of Factory Inspector visiting Union Carbide (1980)

This 1980 letter from the Factory Inspector in Bhopal demonstrates the laxity of the "culture of safety" at the Bhopal pesticide plant, and the generally unconcerned attitude of corporate officers about chemical injury.

Inspector's or Certifying Surgeon's Remarks:-

Union Carbide India Ltd, Bhopal
Dated 2.8.80

Visited the factory today at work along with Shri. M.C. Jain, Asst. Director of Industrial Health and Safety, Bhopal. Shri Pande, Asst. Industrial Relations Manager took us round the plant

Air Compressor Section: Identification mark and date next test and examination.

N2-bottle be painted.

Water drain of this tank be operated regularly.

Plans of workman dining and lockers room were scrutinized on the spot. There are in addition to [illegible] and rest room.

Enquired into the accident case of Mr. A.N. Banerjee, draftsman, who got the burn injury when he was supervising and standing nearby at the time of the drum was being cut by the welder. It was learnt latter on that the drum was containing some volatile chemical after decontamination. The management is hereby directed to ensure that the cutting of such drums shall be carried out only after ascertaining that the drum has no chemical.

Also enquired into the accident case of minor nature since last inspection.

MIC and Co. plant were down at the time of inspection.

By Director
Indl. Health & Safety, Bhopal

Source: Sambhavna Documentation Center, Bhopal.

Correspondence between the Bhopal Worker's Union and the Union Carbide Corporation (1982)

Letter by M.L. Ranjhi regarding the death of worker Mr. Ashraf Mohd. from chemical exposure, with a response from Union Carbide.

Union Carbide Karmachari Sangh
BHOPAL (MP)

Date : 9.2.82

To,
Mr. R. Oldford
President U. C. C. (A. P. D)

Sub : Safety in plant's operation at APD UCIL Bhopal

Sir,

As per subject matter we hereby bring to your notice the loss of life of Mr. Ashraf Mohd. Khan while working in MIC plant on 25 December 1981. Furthermore numerous other accidents have occurred in the Plants due to lack of proper check and replacement of [two words illegible here] or corroded equipment.

The average worker is exposed to toxic and hazardous chemicals without any coverage or compensation for the same. This exposure is harmful and corrective measures may be initiated by your office.

We request you to review our proposal seriously and implement corrective measures.

Yours faithfully,

(M L Ranjhi)
Working President
For U.C.K.S. Bhopal

UNION CARBIDE INDIA LIMITED
KALI PARADE, BERASIA ROAD, BHOPAL 462 018
TELEGRAM: UNION CARBIDE, BHOPAL TELEX: 705-204

D N CHAKRAVARTY
GENERAL WORKS MANAGER

March 24, 1982

Mr. M L Ranji
Working President
Union Carbide Karmachari Sangh
BHOPAL

Dear Sir,

This has reference to your letter dated February 9, 1982, personally delivered by you to Mr. R Oldford, President, Union Carbide Agricultural Products company Inc. regarding the accidental death of Mr. Ashraf Mohd. Khan.

We wish to impress upon you that the Management is aware of its responsibilities to provide protection to its employees and has, therefore, imparted technical skills of a very high order through intensive in-house training program that required all employees to fully understand the safety aspects of every job and every chemical. Sophisticated safety wears have been provided to ensure total safety of its employees. Besides this, the workmen are trained in well established procedures for handling, manufacturing and maintenance of equipments for toxic gases which include knowledge on hazardous job permit, master card locking and tagging, vessel entry, testing inside and outside and total familiarization on the use of protective and respiratory equipments.

These measures are meant for prevention against accidents and exposure to toxic gas.

It is a matter of regret that because of unmindful removal of safety mask before decontaminating himself, Mr. Ashraf Mohd. Khan suffered exposure to toxic gas. It is to be noted that all other employees in the adjacent area who had taken prescribed safety procedures did not have any ill effects.

We wish to reiterate that the Company is firmly committed to maintain safe working conditions in the Plant and ensure its adherence. Your allegations about occurrence of numerous accidents is incorrect.

As already communicated to you, the Management has announced an ex-gratia payment of Rs.50, 000/- besides all other legal dues, to the widow and family of the deceased workman.

We would call upon you to impress all our workmen to follow the safety rules without fail which will prevent recurrence of any accident.

Very truly yours,
D. N. Chakravarty
GENERAL WORKS MANAGER

Source: Sambhavna Documentation Center, translated from Hindi, 2005.

Bhopal, sitting on the edge of a volcano... (1982)

Rajkumar Keswani, a journalist concerned about the state of the factory, wrote a series of articles in the Rapat Weekly, begging the city to take notice of the dangers the Carbide factory posed. He concludes "For now Bhopal sleeps, till the next morning and possibly to never get up some morning."

Rapat Weekly (Issue 2 year 5 Bhopal Friday 1 October 1982 60 paise)

BHOPAL SITTING AT THE EDGE OF A VOLCANO

Bhopal a pile of dead humans in one to one and half hour

Wake up, people of Bhopal, sitting at the edge of a volcano! No savior will save you from this foreign death! Those who used to talk about our rights have pawned their voice and those who could relieve us from this misery are themselves in chains in golden cages that were made for parrots.

Union Carbide's Phosgene storage tank laughs at the fate of Bhopal and Bhopal is lost in Nero's flute.

Union Carbide established the demon MIC plant in 1980 after obtaining its license in 1977 and it has played with many lives since then. Some time back 24 people were hanging between life and death due to Phosgene leak and after suffering for several months are back fighting death to earn their bread.

In 1967–68 when Union Carbide was established, the site of the factory was outside the municipal limits. The area later was brought within the municipal limits and in 1975 the then Administrator of the Municipal Corporation Mr. Mahesh Neelkanth Buch served a notice on Carbide asking them to relocate their factory outside the municipal limits. Before Mr. Buch could take any steps, because of the strong "fate line" of Carbide he was transferred and the matter was buried and at the same time as per a deal, the General Manager of Carbide, at the time Mr. C. S. Ram, donated a sum of Rs.25,000 to build the Vardhaman Park. Thus the notice was sent to the freezer.

This matter dates back to 1977 when the erection of the MIC plant began and to 1980 when the plant started operations. Prior to this, this chemical used to come from the corporation's plant in the USA.

Accidents began in the plant from the time it started. Many accidents were covered up. On December 26, 1981 Mohammed Ashraf was killed due to leakage of Phosgene gas while he was working in the Carbon Monoxide plant. This situation continued and in January '82, phosgene leaked once again during the visit of the President of the Agricultural Products Division of Union Carbide from USA that led to acute sickness of 24 persons who had to be hospitalized and for many months the workers suffered.

Right after they got out of the hospital the workers went back to their fight with death.

In our last issue we had informed about the threat posed by this factory to the entire city and the hazards for the workers. Phosgene gas that was used by Hitler in his gas chambers, and that is used for the production of methyl isocyanate, is stored in a tank in this factory and if that leaks or explodes it will take one to one and half hour for the death of the entire population of this city.

The dangers of phosgene are described in the Phosgene Unit Operating Manual of Union Carbide that has been written by M/s P. K. Behl, S. P. Chowdhary, C. R. Aiyar and S. Khanna and reviewed by Mr. K. D. Ballal.

The four feet wide six and half feet long phosgene storage tank built with steel contains 460 gallons of gas at one time. According to experts, it is quite possible that the gas is corroding the tank from inside and there is no way to know. The city will perish for sure and there is no power that can save the people of this city.

The above manual states that this gas dissolves in air in a few seconds and ordinary people cannot recognize this. It smells of freshly cut hay.

While this threat on human lives continues to loom over us our leaders, officials, our government and other people choose to be silent. But the silence has its reason. Union Carbide pays for this silence.

According to reliable sources Union Carbide has made a room of their Guest House in Shyamla Hills available to the Chief Minister on a permanent basis where he carries out his personal affairs. Usually he arrives there between 11 o'clock and 1 o'clock but he goes there at different hours too. Under such circumstances, what has Carbide to fear?

There is an opposition party called Bharatiya Janata Party, whose leaders do not tire of talking of morality and principles but keep quiet when this matter comes to the fore. The reason is selfishness. In Union Carbide the Bharatiya Janata Party's worker's organization – Bharatiya Mazdoor Sangh is sitting in the lap of the management and the Party too is heavily benefited by the company. This has also come from the mouths of their responsible officials and what else could be the reason for silence on this matter?

There is one hospital in this city – Hamidia Hospital. The senior doctors of this hospital are like the moon and the stars to reach whom you need an Apollo or lunar Module. Those with Apollos reach them and thousands and thousands of patients can only see them in the manner prescribed by Yashoda* – in their reflection in a water filled pot and die in the hope of seeing them. For these moons and stars Carbide not only has the Apollo it recently gifted two imported machines to this hospital. Given this why won't they choose to be silent?

This is an awful conspiracy against humanity. People who are busy with their own affairs are unmindful. Those who know, remain silent. Death is creeping in. For now Bhopal sleeps, till the next morning and possibly to never get up some morning.

Published by Rajkumar Keswani, Printed at Bhoomika Printers, Hawamahal Road, Bhopal. Editor, Rajkumar Keswani Telephone 76156

* Yashoda: mother of Krishna, who advised the 16,000 consorts of her son (the "divine lover") to satisfy themselves with his reflection in their water pots. (Note and article translation, 2005, by Satinath Sarangi.)

Source: Sambhavna Documentation Center, translated from Hindi, 2005.

Letter to the Chief Minister of Madhya Pradesh from Rajkumar Keswani (1982)

In this poetic letter to Arjun Singh, then chief minister of the state, Keswani, a journalist and concerned citizen, begs the minister to investigate the Carbide factory before Bhopal "turns into Hitler's gas chamber."

To,

Mr. Arjun Singh
Chief Minister, M. P. Government,
Bhopal.

Respected Sir,

This letter reaches you carrying many of my complaints. But these complaints are not mine alone; they spring from the miseries of an entire community.

Along with this letter, I am sending you copies of three issues of the weekly "Rapat" [Report]. The primary concern of these issues [of Rapat] are the death threats to the city by Union Carbide, and the sobs of the workers of "Monocaster" section that have been buried between the machines and the files of the Government Press. There is other news enclosed about corruption; I give that second priority.

These are the motives for writing this letter; and if you can understand Namdev Dhasal's anger* in the context of his social oppression and perceive it as justified, then you will not view my anger now out of context. I know for a fact that the city of Bhopal is in danger; save it. Listen to what this crazy person has to say and get it investigated – then you will believe it.

I am well aware that my writing and my newspaper do not amount to much. They will not bring about any revolution; no one in power will lose their throne – but should I keep silent just because of that? From the fear that I will be labeled a yellow journalist, should I be blind [to these dangers]? No, I will not give up, I will fight with the firm determination – I will not let this city turn in to Hitler's gas chamber.

There is lot to say. Lots. But at present there is only one important thing, be it just a peep; but please take a look and see what death has in store for the living in Bhopal.

I know you can find an answer.

On my own tune,

Rajkumar Keswani
Bhopal

Date: 15-10-82

Enclosures: As above.

* Translator's note: A contemporary "Dalit" [low caste] poet, well known for angry non-rhyming poems with large component of expletives.

Source: Sambhavna Documentation Center, translated from Hindi, 2005.

Citizen's letter to Union Carbide (1983)

A letter to Union Carbide India Limited by Shahanawaz Khan, Advocate, warning of the dangers of the Bhopal factory.

Shahanawaz Khan (B. A., LL.B)
Advocate
Jehangirabad, Near Amwali Masjid
BHOPAL
Office: Chhola Road, Bhopal

Date: 4.3.83

Registered AD

To,
Union Carbide India Limited
Through General Manager
Kali Parade, Bhopal

Sir,

As a responsible citizen of Bhopal, as a member of the District Executive Committee of the Communist Party of India, and as a lawyer, I am sending this notice to you:

1. That your factory is situated in the midst of dense residential area within the city limits of Bhopal. About 50,000 citizens reside in Chhola, Nishatpura, Jaiprakash Nagar Colony, Kajik Camp Kabadkhana, Timber Market, Chhola Kainchi, Teela Jamalpura, Krishna Colony, Firdous Nagar and other communities in the vicinity of your factory.

2. That several kinds of poisonous gases and inflammable and poisonous liquids are used in your factory. Phosgene, MIC, carbon monoxide are some of the poisonous gases, the leakage of which can cause the death of any citizen. Right in front of Jaiprakash Nagar there are caustic tanks in your factory next to the road. These tanks contain caustic that can lead to loss of thousands of lives if they burst or catch fire.

3. That earlier there have been accidents of this kind in your factory in one of which a person has lost his life. Some time back there has been a terrible accident in your factory.

4. That waste water that contains poisons flows through a pipe line going through the middle of Jaiprakash Nagar Colony and falls into the sewage canal near the crematorium. This poisonous water poses danger for the residents and cattle. Poisonous water going into the earth from this pipe

line and traveling through aquifers is contaminating the wells in Jaiprakash Nagar and hand pumps and tube wells in the area. Because of this the well water in this area cannot be used for drinking.

5. That periodic leakage of gases used in your factory have their effect on the communities adjacent to the factory. Due to these effects residents in this area occasionally suffer from breathlessness, burning in the eyes and watering of eyes.

6. That the lives of the residents of the area are in danger because of the routine use of poisonous gases and poisonous and inflammable liquids in your factory. There is an atmosphere of fear and there is always the danger of an untoward incident. The lives of 50, 000 people are in constant danger and the spectre of death looms over them.

Therefore through this notice you are being directed to stop the use of poisonous gases and poisonous liquids and inflammable substances within 15 days of receiving this notice or else I will be forced to take legal action in a competent court of law against your factory, the full responsibility of which will be on Union Carbide.

Please note that a copy of this notice is kept safely in my office.

Yours sincerely,
Shahanawaz Khan
Advocate

Source: Sambhavna Documentation Center, Bhopal.

Carbide's internal documents (1972–1984)

Internal documents of the Union Carbide Corporation describing the use of unproven technology in the Bhopal plant, and the differences between it and its "sister" in Institute, West Virginia, USA.

Re.: Technology Risks

"The comparative risk of poor performance and of consequent need for further investment to correct it, is considerably higher in the UCIL operation than it would be had proven technology been followed throughout. CO and I-naphthol processes have not been tried commercially and even the MIC-to-Sevin process, as developed by UCC, has had only a limited trial run. Furthermore, while similar waste streams have been handled elsewhere, this particular combination of materials to be disposed off is new and, accordingly, affords further chance for difficulty. In short, it can be expected that there will be interruptions in operations and delays in reaching capacity or product quality that might have been avoided by adoption of proven technology." UCC 04206 – para 3

"Rationale for EIA ratings on the MIC-based pesticide unit to be located at Bhopal, India."

"There are some differences in basic design between the Institute, West Virginia, and Bhopal facilities." UCC 04295 – para 3

"C. Methyl Isocyanate"

"The Bhopal facility has been given 1.5 ratings because of expected Naphthol emissions from the solar ponds and chloroform emissions from a vent scrubber. Institute has no solar pond, and the corresponding chloroform discharge is flared. The EIA (Environment Impact Assessment) ratings in this case for Institute, were zero." UCC 04296

The Implications of Carbide's discovery documents (2002)

Lawyer for Bhopal Survivors and Activists, H. Rajan Sharma, explains the significance of the Carbide discovery documents showing that design flaws may have caused the Bhopal disaster.

The 1973 documents, obtained through the discovery, show that UCC, in order to retain control of its Indian Subsidiary, decided to reduce the amount of its investment in the Bhopal plant from $28 million to $20.6 million, and that this meant the use of unproven technologies. Why was it necessary for UCC to reduce investment in order to retain control of UCIL?

The Indian government had passed the FERA [Foreign Exchange Regulation Act], which required the dilution of foreign equity from 60 per cent to 40 per cent in order to promote import substitution as an economic development strategy. The only way to avoid equity dilution was to back-integrate the formulation unit into a manufacturing plant, that is, to transfer technology for making "Sevin" to India (instead of simply formulating it). However, FERA also stipulated that at least 25 per cent of all equity for such back-integration would have to be raised domestically in India, which would have meant a lower equity percentage for Carbide. Carbide wanted to retain at least 51 per cent control; therefore, it decided to reduce the cost of investment overall, so it could retain at least 51 per cent equity in UCIL. The way this was done was to transfer substandard technology, use cheap materials and cut corners. Hence the reduction in overall investment from $28 million to $20 million. It also decided to use a process that had only a "limited run" because it was cheaper. Carbide didn't really want to back-integrate the Bhopal plant or transfer additional technology to India, but

India's FERA laws required it either to do so or to dilute its ownership; so it transferred substandard, inferior and dangerous technology instead.

The 1973 documents show that the Sevin pesticide production system at Bhopal had only a limited trial run. What was the ideal trial run that the system should have been exposed to?

Carbide's contract with UCIL and the Indian government required it to transfer "state of the art" technology, which means proven, established and reliable technology that it was using at the American plant to produce MIC [methyl isocyanate] and Sevin. It is not a question of how long the trial run should have been. Instead, Carbide lied to the Indian government and transferred technology that was both unproven and a process for producing sevin from MIC, which had only been tried on a limited, probably experimental basis, instead of "state of the art" technology.

How do the documents establish that UCC was responsible for the design and running of the plant in Bhopal?

I think they clearly show that UCC controlled the nature of the technology that was transferred and that UCIL officials were substantially dependent on Carbide for operational details and other guidance. There is no specific document that shows this, but the overall history submitted in my narrative affidavit to the court establishes the degree of Carbide's domination of UCIL.

Do the documents only reiterate what we knew earlier about UCC's role in causing the disaster?

I think the proof that Carbide transferred unproven technology for the manufacture of this highly dangerous and volatile substance is probably the most significant development in the history of the Bhopal litigation. Many groups, studies, and the Indian government have claimed or alleged this. Now, for the first time, we have documentary proof from Carbide itself.

Source: "Proof from Carbide Itself." Interview with Himanshu Rajan Sharma, *Frontline, December 21, 2002–January 3, 2003.*

3

Explanations

How did the disaster happen?

WITHIN COMPLEX SYSTEMS, big accidents are usually the result of a series of small actions or omissions. The exact series of events that caused the Bhopal disaster are still disputed, but basically only by the Union Carbide Corporation and the companies it has hired to investigate the event. All other investigations have roughly concurred in their findings. Even Dow Chemical, who has fully owned Union Carbide since 2001 says as little about the disaster as possible. When pressed, John Musser, Dow spokesman, noted that "we don't know what happened in Bhopal, but we believe what Union Carbide tells us."

But the important fact about the conditions leading to the accident are set out in chapter 2: Predictions. That is, there were so many small things wrong, and going wrong, with the factory, that any combination of events at any time could have set off a reaction similar to the one that occurred. The conditions were preexisting; it was the responsibility of the corporation to deal with them. Even if, as Carbide still weakly asserts, the disaster was caused by an incidence of sabotage, that would not make the corporation any less liable.

As Union Carbide spokesman Tomm Sprick wrote to Bridget Hanna [one of the editors] in 2004, "... the facts show – based on several investigations – that the safety systems in place could not have prevented a chemical reaction of this magnitude from causing a leak. In designing the plant's safety systems, a chemical reaction of this magnitude was not factored in for two reasons: the tank's gas storage system was designed to automatically prevent such a large amount of water from being inadvertently introduced into the system; and the process safety systems were in place and operational and would have prevented water from entering the tank by accident." Other investigations cited here show clearly that the process safety systems were *not* in place and operational; and more critically, that there were serious design differences between the plant in Bhopal, India and that at Institute, West Virginia, USA. Interestingly, Union Carbide sees nothing wrong with having designed a system that "could not have prevented a chemical reaction of this magnitude." They didn't plan for it, they argue, so it's not their fault.

The reaction *did* happen accidentally; the safety systems were *not* working; *there was no saboteur*. But there are reasons why this particular factory was built in Bhopal, and in the old part of town. There are reasons why it was built near the dwellings of the poorest people in Bhopal. And there are very specific and scientific reasons why the reaction and the gas leak happened as they did. These reasons are the subject of this chapter.

Beginning with two articles about why the disaster occurred in Bhopal, this chapter proceeds to excerpt from the primary investigations that were done into the causes of the release, including the Arthur D. Little (ADL) report on the disaster commissioned by Union Carbide. Although judgment on the relative veracity of these reports is subjective, it is worth noting that while both the trade union and the Indian Council for Scientific and Industrial Research (CSIR) have based their findings on investigative science, the conclusions of the ADL report turn on an interpretation of the testimony of a "tea boy," tracked down in Nepal, who claimed to have served the Carbide workers on the night in question.

The chapter concludes with the testimony of the worker accused but never formally named by Carbide as the saboteur, the "disgruntled employee" who they say caused the disaster.

And the poor get gassed (1987)

Why and how did the gas disaster happen in Bhopal? An article by New York University media studies professor Arvind Rajagopal.

Madhya Pradesh is economically one of the most backward states in India, and this was where the world's worst industrial disaster occurred. The resources of the most advanced cities in the world would have been strained if they had suddenly had to become welfare apparatuses for more than half their population. The capital of Madhya Pradesh was scarcely equipped to handle it. On the one hand, the State was overwhelmed by the fantastic scale of the disaster, and demonstrated a confusion and chaos that was actually out of proportion to the inadequate infrastructure. One writer observed that, on the night of the accident, "Civil authority was virtually nonexistent." Indeed, there are numerous reports that many of the top officials, including the Chief Minister of Madhya Pradesh, actually fled. On the other hand, the government was determined to maintain the opposite – that everything was under control, that nothing greatly untoward had happened, and that there were no long-lasting effects from the gas leak. To maintain this illusion, if for no other reason, it was necessary to suppress all information sources, and to be fanatical about secrecy at every level. To admit to the extent of the crisis would have meant simultaneously to admit to an incapacity to deal with it, and this could launch a political debacle in place of the industrial disaster being denied. It is not necessary to argue intention here: in the absence of sustained political pressure to change its course, there was no need for the State to overhaul itself and become a massive welfare organization. The gas disaster overwhelmingly affected the poor and marginal inhabitants of the city, who lived in squatter settlements, and worked, for the most part, as contract laborers and hawkers. For a variety of reasons, principally their own physical, economic, and political weakness, they were never able to muster the force to have their demands answered except in fragmented measure.

Additionally, industrialization is, especially to an economically backward state, an important priority. While the basic infrastructures remain undeveloped, a partial sort of development takes place through the encouragement of businesses to invest in plants and factories. The regulatory apparatus for such industries may be virtually nonexistent (as was the case with Union Carbide in Bhopal), and the actual gains from the industry may be limited, but the promised land of industrial progress is a land from which no government wants to be excluded, even if it is not clear about just how it ought to get there. The social costs of industrial enterprises are seen as irrelevant or as eminently affordable, since these are invariably paid by the poor. "Reproduction" of labor is not the issue, since, in the eyes of the planners and policy-makers, there is already a huge redundancy of labor. Killing off several thousand, and maiming a few hundred thousand more, is not something they could object to in principle, as long the victims are appropriately selected.* In Bhopal, with almost miraculous precision, the victims were poor and illiterate. Two points have been made: that the State underplayed the disaster to minimize the political fallout; and that industrial accidents were seen as the price to be paid for "Progress".** The first point indicates that the State was necessarily inhibited in its reaction to Union Carbide – the requirement of maintaining a stratified and unequal society reduced its bargaining strength. The second point indicates that, given its vested interest in a particular kind of development, the State did not see Bhopal as precipitating a crisis in its relations with multinationals, or as highlighting crucial contradictions between industrialization and its social and ecological consequences. Rather, it was seen as an accident for which economic compensation should be sought.

There was a third, related, point. Given its commitment to industrialization, the State seeks to generate financial and technological resources to undertake this development. The State is unable to mobilize sufficient resources internally, and is therefore compelled to seek foreign assistance, principally from multinationals. Thus the State cannot react too strongly against Carbide, the management of the Bhopal case therefore receives extremely low political priority. In a situation where Third World countries compete with each other for the attention of multinationals, strict regulatory action could be seen as part of an unfavorable investment climate. In all of this, the deep ideological commitment to "modernization," and to emulating the example of the West, cannot be underestimated.

The Disaster and Its Causes

Union Carbide Corporation (UCC), in 1984, was the 35th largest industrial company in the U.S. with plants in 38 countries. Of UCC's 1984 assets of $11 billion and sales of $9.5 billion, over 14% derived from holdings in Asia, Africa, the Middle East and Latin America, as well as nearly 22% of its $323 million in profits. At the time, UCC manufactured a wide range of products, from consumer goods to industrial chemicals to pesticides (since its 1986 restructuring, it has sold its consumer products division). In India, UCC operates through its subsidiary, Union Carbide India Ltd. (UCIL), of which it owns 50.9% (the rest is owned by Indian stockholders, including the Government of India). UCIL was formed in 1905, and by 1983 was among India's 40 biggest industrial groups. In 1983, it had sales of $202 million and profits of $8.8 million. In 1969, it opened its pesticides plant in Bhopal. Originally a formulation plant to mix and package pesticides imported from the US, it was expanded, and by 1980 manufactured "Sevin" and "Temik" pesticides from methyl isocyanate (MIC). Union Carbide's 1976 manual, "Methyl Isocyanate," begins by stating that MIC is "reactive, toxic, volatile, and flammable." It is five times more toxic than phosgene, which was used extensively in World War I. The manual warns that MIC is so unstable that traces of impurities, including water, chlorine, and iron, can set off uncontrollable reactions.

On the night of 2 December 1984 the MIC production unit had been shut down for six weeks due to an oversupply of pesticides. At about 9:30 p.m., a new worker, on instructions from a new supervisor, began flushing with water the lines of four process filters which were part of the MIC production unit. These lines were connected to another pipeline (the Relief Valve Vent Header, RVVH) designed to carry toxic gasses escaping from the MIC tanks in the event of a pressure buildup. With notoriously leaky valves, and a newly-installed jumper line (in lieu of a more expensive backup system), paralleling the RVVH in function but without any valves blocking entry in to the MIC tank, the entry route into the tank was laid out.

The temperature in MIC tank 610 was well above safe limits. As an economy measure, the refrigeration unit had been shut down for several months. On 2 December, tank temperatures ranged from 15 to 20 °C instead of the prescribed 0 to 5 °C. The water entering tank 610 reacted with the MIC, excess chloroform (used as a solvent), and free iron (from corrosion of the tank walls), in an exothermic reaction. At about 12:30 am, a mixture of MIC and other toxic chemicals burst past the tank's rupture disk, shot through the RVVH, past the vent gas scrubber, and into the atmosphere.

The vent gas scrubber, the plant's main line of defense against escaping gas, had been on "standby" because the MIC unit was not in operation. In any case, its capacity was only a fraction of what was escaping from the tank. Another safety device was the flare tower, designed to burn off escaping toxins. But a section of the pipe connecting the flare tower to the MIC tank and the vent gas scrubber had been removed for maintenance work. The workers tried to douse the leak with water. The spray reached 100 feet into the air; the gas was escaping from a stack roughly 120 feet high. By the time the reaction ended, about 2 hours later, Union Carbide estimates that of the 90,000 lbs of MIC in tank 610, 54,000 lbs of unreacted MIC and 26,000 lbs of "reaction products" blasted into the atmosphere over Bhopal.

A light, south-southeasterly breeze was blowing, directing the dense cloud toward the city. The effect it had was like a holocaust. The most seriously affected area was located between 5 and 8 km from the factory, though the gasses may have traveled over 30 km, entering all areas of the city, whose population was 800,000. The next morning, bodies were littered all over the streets, and in homes – people killed in their sleep, or killed while trying to flee. The wounded numbered in the hundreds of thousands.

* Dr. Vinod Raina, of Eklavya, an institute for innovative educational action in Bhopal, lectured at several of the city's colleges after the disaster, trying to mobilize student support for citizens' action groups. There was little enthusiasm amongst his audience. At the fifth college he visited, he was confronted by a question from a student in the crowd. Hadn't Malthus already predicted that, once the population expanded beyond its resources, disasters would occur and reduce the population level? So what was all the fuss about? Dr. Raina did not give any more lectures.

** The Director-General of the Council on Scientific and Industrial Research, Dr. S. Varadarajan, said in a press conference in Bhopal on December 15, 1985: "The whole issue has to be seen in the context of the cost-benefit ratio. It has to be seen in terms of necessity. No technological operation is entirely without risks; we can't make advances without them."

Source: Arvind Rajagopal, "And the Poor Get Gassed: Multinational-Aided Development and the State – The Case of Bhopal," *Berkeley Journal of Sociology*, Vol. 32 (1987) (footnotes omitted).

The green revolution and Bhopal's pesticide factory (2002)

Why was a large pesticide factory built in Bhopal in the first place? What were the conditions in India and Madhya Pradesh that motivated Carbide to build the plant, and officials to allow its construction? Ward Morehouse and Chandana Mathur explain.

In one horrifying night and through the long misery of nearly eighteen years of injustice that have followed it, the city of Bhopal in India can be said to have experienced the head-on impact of world historical and political economic processes. In 1969, Union Carbide India Limited (UCIL) set up a plant to manufacture the pesticide Sevin in Bhopal, with the permission and, indeed, the encouragement of the Indian government. This was the heyday of the Green Revolution, introduced in India in the mid-1960s and based on HYV (high-yielding variety) seeds, chemical fertilizers, pesticides, and vastly increased irrigation. The disastrous social and environmental consequences of this technological choice (made by Indian planners, with backing from international agencies such as the World Bank and the Ford and Rockefeller foundations) are well documented, but it is not always remembered that the Bhopal gas tragedy has its roots in the same narrative. The design of the Bhopal plant, as well as the policies relating to its maintenance and operation, were in keeping with the double-standard often applied by transnational corporations in their Third World outposts, where environmental, worker, and community safety issues may be seen as less pressing than at home. Finally, for nearly two decades after the deadly gas leak, the legal struggle to make the Union Carbide Corporation (the parent corporation of UCIL) accountable for the tragedy has played out against the changing backdrop of the demise of the Soviet bloc and the introduction of economic liberalization in India. Even though the new economic reforms were formally put in place in India only in July 1991, the writing on the wall had been quite clear in the years preceding. The Indian State's unwillingness to discourage foreign private investment has been a crucial factor in the continuing injustice in Bhopal.

Source: Chandana Mathur and Ward Morehouse, "Twice Poisoned Bhopal," *International Labor and Working Class History Journal*, #62, Fall 2002.

Causes of the gas release (1985)

A specific analysis of the various causes of the disaster from the report of the international trade union delegation (ICFTU-ICEF) that visited Bhopal in 1985.

The Report of the ICFTU-ICEF Mission to study the Causes and Effects of the Methyl Isocyanate Gas Leak at the Union Carbide Pesticide Plant in Bhopal, India, on December 2-3, 1984

International Confederation of Free Trade Unions International Federation of Chemical, Energy, and General Workers Unions.

This report, based over a site study by a 12 member fact finding committee, is the most accurate and detailed analysis we are likely to have of what happened in Union Carbide's Bhopal Plant and how it came about.

It establishes that while some wrong decisions were made by local plant management, Union Carbide Corporation, also bears a major share of responsibility for the catastrophe.

As for the lesson to be learned here at home, none of the conditions which led to the disaster would have been violations of specific standards or regulations of the Occupational Safety and Health Administration or the Environmental Protection Agency

On an international basis, the ratification and implementation by many governments and employers of International Labor Organization (ILO) conventions on occupational safety and health could prevent such horrible accidents.

Lane Kirkland
President, AFL-CIO

Causes of the Release

MIC storage

Large volumes of MIC were stored in the Bhopal plant. Reports on the volume of MIC in tank 610 at the time of the accident vary from 11,290 gallons (75% capacity) to 13,000 gallons (87% capacity). The low figure is contained in the Union Carbide report, which also maintains that such a volume is "well below the maximum operating level." However, Union Carbide's technical manual on MIC suggests a limit of 50%. Bhopal workers confirmed that all three MIC storage tanks were frequently filled above the recommended level.

According to calculations based on data in the Union Carbide report, tank 611 contained 11,565 gallons of MIC until November 24, after which it was slowly drawn down for production to 5,620 gallons on the night of the accident. Tank 619 also contained a small amount of contaminated MIC, despite the fact that Union Carbide's 1978 Bhopal MIC unit operating manual states that one tank is always to be kept empty in case of emergencies.

This long-term storage of large amounts of MIC was a direct cause of the accident. The accident would not have occurred if the MIC in tank 610 and 611 had been promptly converted to Sevin after the MIC unit was shut down. If either tank 611 and 619 had been empty, it could have been used as a surge tank to contain some of the reacting MIC on the night of the accident, thus giving operators more time to regain control of the reaction. In fact, operators were not even sure how much MIC was in tanks 611 and 619, since many of the gauges in the plant were unreliable and not trusted by the workers. As a result, workers where afraid to open the line to tanks 611 and 619. Since they did not know the causes of the reaction in the tank 610, they feared spreading the problems to the adjacent tanks.

Indeed, it was never necessary to store more than minor amounts of MIC in Bhopal. When the plant was first designed, Edward A. Munoz, the managing director of Union Carbide Ltd., took the position that large volume storage of MIC was contrary to safety and economic considerations. In a sworn affidavit to the Judicial Panel on Multidistrict Litigation considering the Bhopal case, Mr. Munoz said that he had recommended, on behalf of UCIL, that the preliminary design of the Bhopal MIC facility be altered to involve only token storage in small individual containers, instead of large bulk storage tanks. However , UCIL was overruled by the parent corporation, which insisted on a design similar to UCC Institute, West Virginia plant. Other UCC facilities which use MIC (but do not produce it) store the chemical in small containers. Such storage is considered safer, owing to much smaller quantity in each container.

In fact, Union Carbide could have produced Sevin in Bhopal without any MIC storage. Since MIC is a chemical intermediate, the process could have been designed in a such a way that MIC was consumed immediately after it was produced. DuPont is currently building such a plant in Laporte, Texas. DuPont has stated that no more than 20 pounds of MIC will be in the system at any one time. A similar process is currently used by Mitsubishi in Japan.

It should also be noted that Carbaryl, the pesticide produced at Bhopal under the trade name Sevin can be made without MIC. Although that process, like the MIC process, utilizes phosgene, another highly toxic chemical, UCC may have chosen the MIC process on economic grounds. MIC is a chemical intermediate for a number of pesticides, and before the accident Union Carbide's Institute plant sold large amounts of it to other companies. UCC probably had similar plans for the Bhopal plant when it was designed.

Safety Systems

The Bhopal plant had four major safety systems designed to prevent or neutralize an uncontrolled MIC reaction:

- A 30 ton refrigeration unit to cool stored MIC, in order to prevent it from vaporizing or reacting;
- A vent gas scrubber (VGS) to neutralize toxic gases with caustic soda in the event of a release;
- A flare tower to burn vented gases from the MIC storage tanks and other equipment; and
- A water spray system to knock down escaping vapors.

At the time of the accident, according to the workers, three of these systems were not operating.

The 30 ton refrigeration unit had been shut down since June 1984. There were no mechanical problems with the system; it was taken out of service to save money. The Freon refrigerant had been drained out for use elsewhere in the plant. The shutdown was in violation of established operating procedures.

The vent gas scrubber (VGS) was turned off in October, 1984, apparently because the supervisors thought it was not necessary when MIC was only being stored and not produced. In addition, the caustic flow indicator was malfunctioning, so it would have been difficult to verify whether the unit was operating or not.

The flare tower had also been out of service since mid-October. A section of corroded pipe leading to it had been removed even though replacement pipe was not ready. The workers stated that the replacement pipe could have been prepared in the plant, and should have taken only four hours to install. However, as of 2 December, the pipe had not been replaced, and the escaping gas could not be directed to the tower. In addition, the company had compromised the reliability of the flare tower even before it was disconnected. The tower was originally built with a backup set to fuel gas cylinders to ensure that the pilot light stayed on. However, the backup system was discontinued to save money.

In spite of all of this, even if the systems had been operating, it does not appear that they could have contained the massive release of MIC gas.

There is some controversy over whether the vent gas scrubber operated at the time of the accident. But if the figures released by Union Carbide are correct, the gas escaped at rate of 400–800 lbs/minute, at a pressure and temperature approaching 200 psig and 200 °C. Union Carbide's 1978 Bhopal Operating Manual gives the "maximum allowable working pressure" of the VGS as 15 psig at 120 °C, and the "nominal feed rate" as 3.2 lbs/minute. Indeed the manual lists "high pressure in the vent scrubber" as a process "upset," for which the remedy is "check for the source of release and rectify." Similarly, "toxic gas release to the atmosphere" from the VGS is also considered as "upset," caused by "high release of the toxic streams from the process," for which the remedy is "check the source of the release and normalize." In short, the VGS was never designed to handle the kind of release which occurred on December 3, and probably could not have significantly reduced the severity of the accident, even if it had operated.

The MIC unit operating manual does not contain the details of the flare tower, and it is unclear whether it was capable of safety flaring MIC at the rate it was being released. However, some workers believe that had the flare tower been put into operation during the release, the enormous MIC cloud near the tower would have exploded, destroying piping systems in the plant and releasing even more MIC.

According to workers interviewed, the water spray shroud which was activated the night of the accident did not reach the level of the gas release, and was therefore useless. In 1982, Union Carbide Corporation, after inspecting the Bhopal facility, had recommended a new, larger water spray system, but it was never installed.

Maintenance

Inadequate maintenance was a long standing complaint at the Bhopal plant. The poor maintenance of the major safety systems has already been described. These problems extended to production equipment.

According to the workers, leaking valves and malfunctioning gauges were common throughout the facility. A 1982 Union Carbide Corporation inspection of the plant by U.S. safety personnel noted such problems, and this resulted in the replacement of valves in the MIC unit; but at the time of the accident, valves and pipes had again corroded and leaking valves were a serious problem. Leaking valves probably allowed water to enter the tank. Broken gauges made it hard for MIC operators to understand what was happening. In particular, the pressure indicator/control, temperature indicator and the level indicator for the MIC storage tanks had been malfunctioning for more than a year.

Manning

At the time of the accident, the Bhopal plant, including the MIC facility, was operating with reduced manpower. According to the workers and published reports, the plant had been losing money, and in 1983 and 1984 there were more personnel reductions in order to cut costs. Some workers were laid off and 150 permanent workers were pooled and assigned to jobs as needed. The workers we interviewed said that employees were often assigned to jobs they were not qualified to do. This practice was also noted by Union Carbide Corporation in its 1982 inspection report. If the workers refused to do the job which they were assigned on grounds they were not trained, their salaries were reduced.

In the MIC facility the production crew had been cut from 12 (11 operators, 1 supervisor) to 6 (5 operators, 1 supervisor), and the maintenance crew reduced from 6 to 2. According to the workers, the maintenance supervisor position on the second and third shifts had been cut on November 26, less than a week before accident. The maintenance supervisor was responsible for preparing all maintenance work and giving instructions for the job. The workers indicated that it would have been the responsibility of the maintenance supervisor to prepare the pipe which was being flushed with water the night of 2 December 1984, including to prevent the entry of water into the pipes leading to tank 610.

When the post of maintenance supervisor was eliminated, these responsibilities apparently shifted to the production supervisor. But according to the workers, the production supervisor on duty the night of 2 December had been transferred from a Union Carbide Battery plant one month before and was not fully familiar with either operating or maintenance procedures.

Training

Training was a major problem at the Bhopal plant. At the time the MIC facility was opened in 1980, 25 people were sent to the United States for training. But due to high turnover – 80% in the MIC plant from 1982–1984 by the workers' estimate – few of the people originally trained in the MIC operation in the U.S. remained in Bhopal.

The workers said that they had been given little or no training about the safety and health hazards of MIC or other toxic substances in the plant; they thought the worst effect of MIC was irritation of the eyes. Even a maintenance worker who had been assigned to the MIC facility since it first began operation in 1980, stated that he had been given virtually no training about the safety and health hazards of MIC.

Language also may have contributed to the lack of understanding about MIC and other hazards. All signs and operating procedures were written in English, even though many of the workers spoke only Hindi. Workers stated that if they wrote in the log books in Hindi, they were reprimanded.

Insufficient Corporate Attention to Safety

In May of 1982, a three man team from Union Carbide Corporation in the USA inspected the Bhopal plant. In its report, the team found a number of "concerns" which they classified as "major" or "less serious," with major concerns being those that represented "either a higher potential for a serious incident or more serious consequences if an incident should occur."

The team listed 10 major concerns. Among them were:

3. Potentials for release of toxic materials in the phosgene/MIC unit and storage areas, either due to equipment failure, operating problems, or maintenance problems.
4. Lack of fixed water spray protection in several areas of the plant.
7. Deficiencies in safety valve and instrument maintenance program.
8. Deficiencies in Master Tag/Lockout procedure application.
10. Problems created by high personnel turnover at the plant, particularly in operations.

Problems like these led to the catastrophic MIC release 2-1/2 years later. But while the team classified these items as "major" in relation to the other "less serious" concerns, the overall message UCC sent its Indian subsidiary was at best confusing. The report's opening summary states:

"The team was very favorably impressed with the number and quality of operating and maintenance procedures that had been developed and implemented in the past 1–2 years. These procedures together with the Job

Safety Analyses detailed for most operations, constitute a major step for all concerned....No situation involving imminent danger or requiring immediate correction were noted during the course of the survey."

In accord with corporate procedures, UCIL prepared an action plan in response to the 1982 inspection and sent periodic progress reports to the United States until June 1984. But UCC never sent a follow-up team to Bhopal in the two-and-half years between the inspection and the accident. The workers report that UCIL did temporarily fix many of the items cited in the 1982 report, but by the time of the accident, conditions had again deteriorated.

Labor Relations and Management Disputes

The Bhopal plant was plagued by labor relations problems and internal management disputes. In India there can be more than one union representing workers at a plant. At Union Carbide Bhopal there had been conflicts between a union affiliated with the Indian National Trade Union Congress (INTUC) and an independent union, over who represented the majority of workers in the plant. Even at the moment of this writing there is a court case pending on representation rights. According to the workers, the management tried to use this inter-union rivalry to its advantage in contract negotiations, to reduce manning levels and in other labor relations matters.

It also appears that there were internal management disputes in the company .The management structure of UCIL changed before the accident, and the Bhopal Pesticides plant was put under the direction of the Union Carbide battery division in India. According to the workers, this resulted in management conflicts within UCIL and the transfer of managers to Bhopal from the battery operation who were not fully trained about the hazards and appropriate operating procedures for the pesticides plant.

Failure to Respond to Previous Accident and Workers Warnings

The December 2-3 1984, MIC release was not the first accident at the Bhopal plant. Published reports and interviews with workers we spoke with indicate that there were at least five chemical accidents in the plant between 1981 and 1984.

In December 1981 a phosgene leak injured three workers; one of the workers died the next day. Two weeks later in January 1982, 24 workers were overcome by another phosgene leak. In February 1982 an MIC leak affected 18 people. In August 1982 a chemical engineer came into contact with liquid MIC resulting in burns over 30% of his body. And in October 1982 a combined MIC, hydrochloric acid and chloroform leak injured three workers in the plant and affected a number of residents of the surrounding neighborhoods.

Since 1976 the two unions representing Bhopal workers had frequently complained to Union Carbide management and the Madhya Pradesh authorities, including the Factory Inspectorate, about safety and health hazards in the plant. Correspondence obtained from the local INTUC affiliated union, and our interviews in Bhopal, demonstrate that both unions consistently raised health and safety issues with management and the government, and warned of grave dangers if the conditions were not corrected.

In a July 1976 letter to the General Manager of the Bhopal plant, the union listed five serious accidents, including one case of chemical burns and one of blindness resulting from separate incidents. The letter stated:

"On reviewing the above incidents one can conclude that the safety measures are inadequate. Despite the instructions from the government before commencement of production that the worker safety should be given top priority, we feel that you have neglected this aspect."

In an April 13, 1982, letter to the Minister of Labour of Madhya Pradesh, the union wrote:

"Our unit is going to celebrate the safety week from the 14th of April, 1982. But the workers would like to inform you that this function is merely a window-display.... We would also like to point out that our unit is manufacturing dangerous chemicals like phosgene, carbon monoxide, methyl isocyanate, BHC, naphtha and temik."

After the October 1982 combined release of MIC, hydrochloric acid and chloroform, which spread into the community, the union printed hundreds of posters (in Hindi) which they distributed throughout the community, warning:

- Beware of Fatal Accidents
- Lives of thousands of workers and citizens in danger because of poisonous gas.
- Spurt of accidents in the factory, safety measures deficient.

But despite these constant warnings by the union, little was done to correct these problems and prevent a potential disaster.

Source: The International Confederation of Free Trade Unions and International Federation of Chemical Energy and General Worker's Union, Trade Union Report on Bhopal, Geneva, Switzerland, 1985.

Report on scientific studies on the factors related to Bhopal toxic gas leakage (1985)

The report of the Indian Council for Scientific and Industrial Research (CSIR) studies the nature of the storage and use of MIC, and the "factors and circumstances which lead to the chemical reaction and the gas leak." The following is an excerpt:

An Analysis of the Event:

The background and circumstances, the study of the properties of the materials, examination of tank residue, experiments on conditions for formation of various chemical entities, critical examination of relevant features of design have all been described in the previous chapters (sic). This provides a basis for outlining the factors which led to the event, which are given in the following paragraphs:

- Methyl isocyanate (MIC) readily undergoes chemical reactions with explosive violence, which produces a large amount of heat, and allows a large portion of stored liquid MIC to vaporize. This is inherent to the nature of the material. Neither the precise conditions under which such runaway reactions could be initiated in MIC nor its manner of prevention are well known.
- A large quantity of MIC was stored in underground tanks which have many inlets and outlets that can permit entry of contaminants which can trigger off explosive reactions.
- The studies described earlier indicated that a reaction was initiated and the temperature rose rapidly. There was no "rate of rise in temperature" alarm to indicate the rising temperature which would have alerted the operator to an early detection of a runaway reaction.
- Any emergency dumping of liquid MIC into the vent gas scrubber (VGS) would not have been feasible because the alkali available in the accumulator is grossly insufficient. Further more, this would also lead to an abnormal temperature rise in the VGS.
- It is conceivable that gaseous MIC would be emitted due to a rapid reaction inside the tank. This gas is expected to be neutralized by circulating alkali solution in the paced section of VGS. This system is also grossly inadequate to handle quantities of vaporized MIC as large as were emitted during the event. Calculations show that even if the normal design load for VGS is taken into consideration the VGS is inadequate to neutralize a discharge of 28 tons of vaporized MIC in about 2 hours.

- Therefore neither the liquid nor the gaseous disposal system was capable of handling the event which occurred on the night of December 2, 1984.
- The relief valve design could not permit free flow of large quantities of gasses, certainly at the level at which they were generated during the event. Thus, the tank contents were subjected to pressure much higher than 40 psig and correspondingly high temperatures.
- From the examination of the tank residue and from the conditions of formation of the residue, it is surmised that the temperature reached in the bulk storage talk may have been around 250 °C. The total energy balance on the tank also indicated that the probable temperatures would be in the range of 200 to 250 °C. Information from the mechanical examination of the tank indicated that the pressures may have reached 11 to 13 kg/cm^2g with the corresponding temperatures in the range of 200 to 350 °C.
- The chemical analysis of the tank residue clearly shows the evidence of entry of approximately 500 kg (1100 lb) of water. The fact that the tank 610 was not under pressure of nitrogen for approximately two months prior to the accident also indicates that conditions existed for entry of contaminants such as metallic impurities through the high-pressure nitrogen line. As emphasized earlier, many such impurities have a catalytic effect on the possible reaction MIC can undergo.
- The hydrolysis of MIC with about 500 kg (1100 lb) of water by itself and in the absence of other contaminants is not expected to lead to thermal runaway conditions. The presence of this quantity of water would have possibly resulted in reaction with about three to four tons of MIC, generation of carbon dioxide, breaking of the rupture disc and release of CO_2 along with small quantities of MIC, 100–150 kg/hr (220–330 lb/hr) would have been released. Such emissions will cease once the water has been consumed.
- The presence of trace amount of metallic contaminants derived from the material of the tank or its attachments or from extraneous sources, may not necessarily initiate a violent reaction under dry conditions. Small amounts of local trimerization may take place, as noticed throughout the pipelines and plant. However, the ingress of water would provide for active species of initiator to be generated and distributed in the liquid.
- This implies that conditions which were ripe for the initiation of a runaway trimerization reaction already existed in tank 610 on the day of the event and that the entry of water would generate active

initiators and the hydrolysis of MIC would provide the necessary heat also for the trimerization reaction to take off with explosive violence. The carbon dioxide evolved upon hydrolysis would provide the necessary mixing leading to even more rapid chemical reactions. The storage tank thus equaled the conditions of a well-mixed tank reactor, supplied with heat.

- Once initiated, the trimerization reaction rapidly led to a temperature increase leading to levels as high as 250 °C, with autocatalytic and auto-thermal features. At these high temperatures, secondary chemical transformation occurred leading to the complex mixture of products actually found in the tank 610 residue.
- 500 kg (1100 lb) of water could have entered the tank 610 either from the MRS condenser (through condenser tube leaks) or through RVVH/PVH lines. The fact that about 50–90 ppm of sodium was found in the tank residues is also significant. The observed sodium levels are substantially higher than what would have normally been present due to the entry of 500 kg (1100 lb) of water (below 0.5 ppm).
- Entry of water from the MRS condenser appears less likely as the MRS make line from MRS condenser to tank 610 was flushed with nitrogen, evacuated and blinds were inserted in the line between 23 to 25 October, 1984. Water could have entered from the MRS prior to 23 October but it would have probably reacted much earlier than 2 December, 1984
- A detailed analysis indicated that water entry through RVVH/PVH lines quite likely. It has been reported that around 9.30 p.m. on 2 December 1984, an operator was clearing a possible choke of the RVVH lines downstream of phosgene stripping still filters by water flushing. Presumably the 6" isolation valve on the RVVH was closed but a slip blind had not been inserted. Under these conditions where the filter lines are choked, water could enter into the RVVH, if the 6" isolation valve had not been tightly shut or passing.

Furthermore, it is likely that alkaline water could have backed up from the VGS accumulation into the RVVH and PVH under certain conditions. Indeed, several liters of alkaline water were drained from the RVVH/PVH lines in the MIC structure in May 1985, lending credence to such a possibility.

- Furthermore, RVVH/PVH lines are made of carbonized steel. Back up water or alkali through these lines increase the possibility of metal contaminates entering the tank, especially in the absence of positive nitrogen pressure.

- The water that entered RVVH at the time of water flushing along with backed up alkali solution from the VGS already present could find its way to into the tank 610 through the RVVH/PVH lines in the blow down DMV or through the SRV and RD. To account for say 60 ppm of sodium in the residue, quantity equivalent to 25 liters (about 6.5 gal) of 5% sodium hydroxide would have been sufficient. Such an eventuality would have required that the blow down DMV was malfunctioning or the SRV was not tightly shut and rupture disc had been damaged.
- The absence of practically any nitrogen pressure on tank 610 for over a month and the observation that MIC was leaking out of a branch of RVVH on the downstream side of Regeneration Gas Cooler Pressure Safety Valve at 23.30 hours on the night of 2 December, 1984 indicated the possibility of malfunctioning of blow down DMV.
- Opportunities for intrusion of water, alkali and metal contaminants into tank 610 thus existed from 22 October, 1984 and into tank 611 as well during 30 November to 1 December, 1984, when there was negligible positive nitrogen pressure in these tanks.
- Trimerization of MIC solid material in small quantities and consequent choking of lines leading to tank was a frequent occurrence and seems to have been well accepted by the plant operating staff. Similarly, cleaning and purging with water of lines associated with the storage tanks was also accepted as a routine procedure. The hazards presented by ingress of water or other contaminants which could cause trimerization and lead to choking was not appreciated and the tank 610 was allowed to stay without positive nitrogen pressure from 22 October to 2 December, 1984.
- The UCC brochure on MIC makes reference to the reaction of water to produce DMU and TMB. The heat generated would be related to the quantity of water. The brochure also mentions metallic contaminants could lead to violent reactions of MIC which has the unique combination of properties of explosive reactivity, ready volatility and high inhalation toxicity. It seems possible that small amounts of water in presence of trace amounts of metallic contaminants could set off explosive reactions and leakage not containable in the inadequate VGS system.
- It was entirely unnecessary to provide facilities for storage of such large amounts of MIC in tanks. The quantities stored were quite disproportionate to the capacity of further conversion of MIC in downstream unit. This permitted MIC to be stored for months together without appreciation of potential hazards.

- The events on 2/3 December 1984 arose primarily from these facilities and accepted practices.
- The rapid rise in temperature and violent reaction that occurred necessitate the onset of metal ion catalyzed polymerization. The presence of chloroform has no influence whatsoever initiating or accelerating the runaway reactions. The chloroform present was involved in chemical transformation when the temperature had risen about 200 °C at which stage all the water would have already been consumed by reactions.
- The quantum of toxic leakage by violent chemical reaction is not related to the amount of metal and water which initiate the reaction but to the quantity of MIC stored in a single container. If 42 tons of MIC had been stored in 210 stainless steel drums of 200 litre capacity each, as an alternative to a single tank, there would be no possibility of leakage of more than one fifth of a tonne and effects of even such a leakage could be minimized by spray of water or alkali.
- It has been reported that a large number of leakages of MIC in comparatively small quantities have occurred from storage vessels and tanks in Union Carbide. The causes of such leakages have not been made known. While Union Carbide product brochure refers to chemical properties and reactivity of MIC including possibilities of violent reaction as well as of possibilities of spillages from drums and tanks during transport, no information is provided on the extraordinarily toxic effects of inhalation. Public preparedness for eventualities arising out of leakage would have been substantially greater if information on these had been generally available. For instance, substantial care is exercised in the storage and transport of explosives or in handling even weakly radioactive materials or in the use of X-ray equipment, due to the awareness of hazards.

In retrospect, it appears the factors that lead to the toxic gas leakage and its heavy toll existed in the unique properties of very high reactivity, of the material in very large size containers for inordinately long periods as well as insufficient caution in design, in choice of materials of construction and in provision of measuring and alarm instruments, together with the inadequate controls on systems of storage and on quality of stored materials as well as lack of necessary facilities for quick effective disposal of material exhibiting instability. These factors contributed to guidelines and practice in operation and maintenance. Thus the combination of conditions for the accident were inherent and extant. A small input of integrated scientific analysis of the chemistry, design and controls relevant to the manufacture would have had an enormously beneficial influence in altering this combination of conditions, and in avoiding or lessening considerably the extent of damage of December, 1984 at Bhopal.

Source: Sambhavna Documentation Center, Bhopal.

Investigation of large-magnitude incidents: Bhopal as a case study (1988)

Union Carbide hired the consulting firm, Arthur D. Little, Inc. to investigate what happened in December 1984. The results of that investigation were presented at a technical meeting in London in May 1988. Excerpts from their report follow:

Psychological Factors

Perhaps because of the enormity of the event, many people, even those only peripherally involved, tended to remember in detail and with great clarity the sequence of events of that night. Nevertheless, people experienced the event in different ways, thus yielding, for example, varying estimates of the duration of the actual release.

The tendency of plant workers to omit facts or distort evidence was also clearly evident after the Bhopal incident, making the collection of evidence a time-consuming process. In investigating any incident in which facts seem to have been omitted or distorted, it is necessary to examine the motives of those involved. The story that had been initially told by the workers was a preferable one from their perspective, because it exonerated everyone, except perhaps the supervisor. According to this version, the reaction happened instantaneously; there was no time to take preventive or remedial measures, and there was no known cause. Without a cause, no blame could be established.

Because critical facts were being deliberately omitted and distorted, the investigation team had to continually review and reanalyze each new piece of evidence and to assess its consistency and veracity with hard evidence and known facts. Ultimately, several firm pieces of evidence came to light – evidence that simply did not fit the story told initially by the workers, and that eventually led to the conclusion that a direct water connection had been found by the workers, but had been covered up.

The Direct-Entry Chronology

The results of this investigation show, with virtual certainty, that the Bhopal incident was caused by the entry of water to the tank through a hose that had been connected directly to the tank. It is equally clear that those most directly involved attempted to obfuscate these events. Nevertheless, the pieces of the puzzle are now firmly in place, and based upon technical and objective evidence, the following sequence of events occurred.

At 10:20 p.m. on the night of the incident, the pressure in Tank 610 was at 2 psig. This is significant because no water could have entered prior to that point; otherwise a reaction would have begun, and the resulting pressure rise would have been noticed. At 10:45 p.m., the shift change occurred. The unit was shut down and it takes at least a half hour for the shift change to be accomplished. During this period, on a cold winter night, the MIC storage area would be completely deserted.

We believe that it was at this point – during the shift change – that a disgruntled operator entered the storage area and hooked up one of the readily available rubber water hoses to Tank 610, with the intention of contaminating and spoiling the tank's contents. It was well known among the plant's operators that water and MIC should not be mixed. He unscrewed the local pressure indicator, which can be easily accomplished by hand, and connected the hose to the tank. The entire operation could be completed within five minutes. Minor incidents of process sabotage by employees had occurred previously at the Bhopal plant, and, indeed, occur from time to time in industrial plants all over the world.

The water and MIC reaction initiated the formation of carbon dioxide which, together with MIC vapors, was carried through the header system and out of the stack of the vent gas scrubber by about 11:30 to 11:45 p.m. Because the "common valve" was in a closed position before the incident and the tank held a strong vacuum when it cooled down after the incident, it is clear that the valve was temporarily opened to permit the entry of water. This also permitted the vapors initially generated to flow (via the PVH) out through the RVVH. It was these vapors that were sensed by workers in the area downwind as the earlier minor MIC leaks. The leak was also sensed by several MIC operators who were sitting downwind of the leak at the time. They reported the leak to the MIC supervisor and began to search for it in the MIC structure. At about midnight, they found what they believed to be the source, viz., a section of open piping located on the second level of the structure near the vent gas scrubber. They fixed a fire hose so that it would spray in that direction and returned to the MIC control room believing that they had successfully contained the MIC leak. Meanwhile, the supervisors went to the plant's main canteen on break.

Shortly after midnight, several MIC operators saw the pressure rise on the gauges in the control room and realized that there was a problem with

Tank 610. They ran to the tank and discovered the water hose connection to the tank. They discussed the alternatives and called the supervisors back from the canteen. They decided upon transferring about one ton of the tank's contents to the Sevin unit as the best method of getting the water out. The major release then occurred. The MIC supervisor called the MIC production manager at home within fifteen minutes of the major release and told him that water had gotten into an MIC tank. (It later took UCC's and GOI's investigating teams, working separately, months to determine scientifically that water entry had been responsible.)

Not knowing if the attempted transfer had exacerbated the incident, or whether they could have otherwise prevented it, or whether they would be blamed for not having notified plant management earlier, those involved decided upon a cover-up. They altered logs that morning and thereafter to disguise their involvement. As is not uncommon in many such incidents, the reflexive tendency to cover up simply took over.

Conclusion

The operators in the MIC unit gave widely contradictory accounts. For example, some stated that the alarm signaling the major release went off only several minutes after tea began at 12:15 a.m., whereas others stated that the tea period in the control room was entirely normal, and they had not noticed anything to be amiss until just a few minutes prior to the major release. The control room operator initially told the media that he noticed the pressure in Tank 610 was 10 psig when the shift began; however, he later stated that the pressure remained at 2 psig until after tea.

Because some of the witnesses directly involved in the incident were initially unavailable for interviews, and because others were rendering obviously contradictory accounts, reports given by the more peripheral figures during the incident became highly important. For these individuals, primarily operators from other units or those who were not present at the time of the incident, there was no motive to distort or omit facts, and their accounts were thus deemed more reliable.

Ultimately, it became clear that the MIC operators knew at least 30 to 45 minutes before the release that something was seriously wrong, and that several had acted in an attempt to forestall the problem.

One of the more reliable accounts came from a witness who had no motive or reason to distort or omit the facts. He was the "tea boy", who served tea in the MIC control room just prior to the major release. With some difficulty, he was located in Nepal, in the Himalayas, and brought to Delhi. Despite the MIC operators' claim of a normal tea period, the tea boy reported that when he entered the unit at about 12:15 a.m., the atmosphere was tense and quiet. Although he attempted to serve tea, the operators refused it. After detailed questioning of scores of operators, it became apparent that those directly involved were unable to give consistent accounts because they were attempting to give very specific details of events that never occurred.

By their nature, large-magnitude incidents present unique problems for investigators. In the case of the Bhopal incident, these problems were compounded by the constraints placed on the Union Carbide investigation team by the Indian Government and, most significantly, by the prohibition of interviews of plant employees for over a year. Had those constraints not been imposed, the actual cause of the incident would have been determined within several months.

Because the investigation was blocked, a popular explanation arose in the media as to the cause of the tragedy. A thorough investigation, which included scores of in-depth witness interviews, a review of thousands of plant logs, tests of valving and piping, hundreds of scientific experiments, and examinations of the plant and its equipment, was ultimately conducted over a year later. That investigation has established that the incident was not caused in the manner popularly reported, but rather was the result of a direct water connection to the tank.

Source: Ashok S. Kalelkar, Arthur D. Little, Inc., Cambridge, Massachusetts, May 1988.

A Carbide worker speaks out (1994)

M.L, Varma, the worker believed to be Carbide's sabotage suspect, tells his story of the night of the gas disaster and of the factory's internal politics.

M. L. Varma age 37
(Carbide Worker Token No. 4557)
Industry Inspector, Government of Madhya Pradesh
(Originally hired for alpha-naphthol plant)

I joined Union Carbide on 28 March 1977 as part of the second batch for the alpha-naphthol plant. I had six months of classroom training and no training on the job. During the on-the-job training period they used us for pre-commissioning and start up of the naphthol plant. I continued working in the naphthol plant but the plant was not running successfully. There were plans for large-scale modifications, for which they shut down the plant. In this period most of the operators of the naphthol plant were transferred to the formulation plant. As there was no qualified operator jobs, we worked as packers and in other labor class jobs. We came back to the naphthol plant when it restarted, but it still did not function properly. They decided to permanently close down the alpha-naphthol plant, so all the operating staff became jobless. They launched a voluntary retirement scheme and about 30 workers resigned. Those of us who remained were sent for theoretical training for the MIC plant. After training, we took exams for selection as an MIC unit Operator. I was selected.

About September 1983, I was sent to the MIC unit for on-the-job training. There they told me that I must learn about the MIC plant from my fellow workers. When the plant was running, it was difficult to take on-the-job training, but somehow I began learning about the MIC process. My demands for assistance were always refused. In this period I was asked to take charge when regular operator was absent from duty. I refused to take charge under these conditions; they had not confirmed me as a regular operator for the MIC plant. I had decided not to take charge until after confirmation because I wanted to be sufficiently trained and I wanted the financial benefits. For these reasons, management refused to confirm me. They said there were no vacancies.

They gave me an oral warning about the job refusal but never gave it in writing since they were aware that I was not confirmed as a regular operator. However there were workers given confirmations who joined UCIL after me and who were less qualified. These persons never resisted management using them in positions for which they were not qualified. For almost a year I refused to take charge as an operator because I was not yet confirmed. Then, in November 1984, MIC plant manager S. P. Choudhry told me that if I did not follow all orders, I would be transferred to other units. I told him that I was not refusing any job for which I was confirmed. I would perform any job for which I was needed if I was trained properly and was receiving proper financial benefits. I told Choudhry that if they were to transfer me due to false charges of job refusal, I wanted it in writing so that I could proceed with a legal response.

After a few days, S. P. Choudhry took an oral test from me about the MIC plant process and said that if I passed the test I would be confirmed. I took the test and was able to give a correct reply to every question. Nonetheless, I was given a failure on the test and told I was not fit for the MIC plant. Then they told me that I would be transferred to the Sevin plant. They mentioned the transfer in their daily notes but did not give me any letter. I argued that a daily note is not sufficient for transfer purposes, or that I required a letter. Nor did they mention the transfer on the notice board.

Even though on the basis of the oral test they said that I was not capable as an MIC plant operator, they had tried months before to force me to take charge in the MIC plant unit when I was not confirmed. The transfer was S. P. Choudhry's way of taking revenge for my not obeying him in the past. Although they announced my transfer on 26 November, I continued to come to the MIC unit. I also began to personally report to the MIC plant superintendent and to the production assistant. This continued until the night of the gas leak. During the week, no action was taken against me for not reporting to the Sevin plant. I went to the MIC unit and sat there because there was no work for me.

On 2 December 1984 I was on night shift. I punched my card around 10:50 and reported to the production assistant of the MIC plant. About 11:15 P.M., I was sitting in the MIC control room along with my fellow workers. Then I went to the tea room at the 200-ton refrigeration unit. Generally, when we are free MIC operators sit in this room. The window of the tea room toward the MIC unit was open. Around 11:30, we felt MIC irritation so we came out from the room to locate the source of the leak. We saw that some water was dropping from the MIC plant structure. Near that water the MIC was in greater concentration. As we came toward the vent gas scrubber side we felt high MIC concentration. We reported the MIC leak to production assistant S. Qurashi. The plant superintendent was also sitting there in the control room. They replied that the MIC plant is down and thus there is no chance of leak. They did not take our report seriously, saying "Koi baht nahin appan chay ke bad dhekhenge" ("Okay, no problem, we'll see after tea").

In the meantime, the tea boy came to the control room and we took tea. Then the plant superintendent went to smoke a cigarette near the security gate, as it is not allowed on the plant premises. Now the time was around 12:30. With the supervisor, we went to the MIC. The operator, Khampariya, was ordered to spray water on the leaking point. The supervisor was not able to trace out the source of the leak. Around 12:50 the leak became vigorous and started coming out from the vent gas scrubber atmospheric line. I was standing in front of the control room when the siren started. After a few minutes, the plant superintendent came back to the MIC unit. As he met me, he asked, "What happened?" I told him MIC was pouring from the top of the vent gas scrubber.

Because of the siren, the emergency squad came to the MIC unit. They tried to control the leak by massive water spraying. I also helped them until the conditions in the area became unbearable. Then along with other workers I left the MIC unit area in the opposite wind direction. The MIC production assistant also fled. When the plant superintendent came back from smoking, he ordered that the loud siren be stopped. This was around 1:00 a.m.. Around 2:00 a.m. when we learned that the toxic release was affecting the communities outside the plant, we argued with the plant superintendent to restart the loud siren. He refused saying it would serve no purpose, but we insisted until he switched it on again. Around 2:15, the gas leak stopped so we returned to the MIC unit and discovered that the MIC production assistant was missing. After some time, we learned that he was lying near the boundary wall. Some workers brought him to the dispensary.

Around 3.00 a.m. I saw many people from outside coming for medical help. Many were in dying condition. A managerial staff member, Roy Choudhry, and others were denying help to these people from outside. We argued with the dispensary staff, telling them that we must provide any help possible since they were affected by a leak from our factory. Finally they began to administer basic first aid. When I came to know that the area in which my family was living was also affected, I rushed home. This was around 5:00 a.m. Outside the plant, I saw how badly the gas had affected people.

I first came to know that UCC was claiming the leak had been caused by sabotage through the newspaper. First they blamed a Sikh terrorist, then a "disgruntled worker." They never mentioned any name of a worker, but gave a detailed description of one who was disgruntled due to being transferred to the Sevin plant but who remained "illegally" in the MIC unit. Immediately, I knew that they were trying to frame me, even though the description did not fully fit. I think this description came after UCC management people and lawyers interviewed ex-workers.

I also gave an interview about my experiences that night. They asked me about my past history with Union Carbide so I told them about my problems regarding confirmation. Then they used this information against me to construct their sabotage theory.

Beyond UCC lies suggesting that I am to blame, there are other reasons why the sabotage theory is clearly incorrect: it is not possible for any worker to put water directly into the MIC tank, as it is a very dangerous job. Further, everyone knows that a MIC and water reaction is very dangerous, not just spoiling the contents of the tank. So, it may be sabotage that caused the leak but not by any worker. If the leak was caused by sabotage, the culprit is the management who was responsible for overseeing the safety of the MIC plant. The leak was a result of continuous negligence, unsafe handling and a poor warning system.

Source: by M. L. Varma, from T. R. Chouhan, *et al.*, *Bhopal: The Inside Story*, New York: The Apex Press and Mapusa, Goa: Other India Press, 1994, revised edition, 2004.

4

Just Compensation?

Civil cases, uncivilized results

THE POTENTIAL FOR civil damages in the wake of the Bhopal gas disaster was legally unprecedented – but that does not account for the degree to which it has failed the survivors. The legal lesson of Bhopal unfortunately is that the bigger the scale of the disaster, the greater the degree of the injustice.

Although both the Indian government and the survivors of the disaster lobbied to have the legal case against Union Carbide Corporation (UCC) tried in the United States – since Carbide was an American company – it was dismissed by Judge Keenan in New York on the debatable grounds that denying India the opportunity to try the case would be a form of continued imperialism – in effect staking the lives of the survivors and the value of the victims on a wager that the Indian judicial system was, or ought to become in the trying of this case, fully independent of corporate and governmental pressures.

In the immediate aftermath of the disaster, the city of Bhopal swarmed with foreign personal injury lawyers, collecting thousands of cases with an eye towards taking a cut of a settlement that it was presumed would be as unprecedented as the disaster itself. Partly as a result of this phenomenon, the Indian government passed the controversial Bhopal Gas Leak Disaster (Processing of Claims) Act, known as the Bhopal Act.

This Act, essentially categorizing the gas affected as juridically incompetent under the edict of *parens patriae*, allowed the government to settle the case out of court with Carbide, with no survivor input for $470 million in 1989 (once disbursed several years later, 95 percent of those who received anything got less than $500 and the average payment was between $280 and $330). Given that the Indian government owned about a quarter of Union Carbide India's stock, this essentially gave them *carte blanche*. Behind closed doors, the officials acted as judge, plaintiff and accused.

This story of collusion was repeated and multiplied, unfortunately, as the compensation money made its way down from the courts into the hands of the survivors. In a perfect illustration of Hannah Arendt's "banality of evil," armies of regular people; doctors, bureaucrats, bankers, judges, middlemen and politicians, blithely conspired to create a compensation and claims system, that, as illustrated by the testimonies and the Amnesty International report below, was not only distributing a paltry sum of money (the interest of which, in fact, the government then held onto until 2004), but that was distributing it in an exploitative, ineffective, unjust, humiliating, and delayed manner. Entire systems of documentation, classification and bureaucracy were invented to regulate the compensation process and subjects, but despite these reams of paper, the only kind that really mattered was legal tender, paid as bribes to judges or middlemen. It was the poorest people, lowest on the class and social strata, who were most affected by the gas in Bhopal. The compensation courts made certain that they stayed that way.

That said, most of the direct, and indeed radical decisions that had the potential to transform this dynamic, came from the lower courts in Bhopal. It was the local courts that gave a decision for interim relief – which would have allowed a real trial against Carbide, rather than a hasty settlement, to occur. It was the Supreme Court that struck it down. Although Judge Keenan in New York told Carbide that they must abide by the decisions of the Indian courts if they were not to be tried in the US, when the local courts in Bhopal decided to reinstate the criminal case against Union Carbide, Carbide, and Keenan, assiduously ignored it. Most recently, the local court in Bhopal has ordered Dow Chemical in India, to present the absconder corporation Union Carbide in the dock. It is to be hoped that a higher court will not munificently decide that Dow really ought not be asked to prevail upon on its fully owned subsidiary to do so.

This chapter begins with the story of a gas widow's struggle to get compensation. Next, Veena Das of Delhi University outlines the major events in the civil case and the Indian government's major legislative response to the disaster. As a result of the exclusion of the survivors from any judicial process, and as a result of the settlement that many viewed as a miscarriage of justice, Bhopal was the subject of a series of sessions held by the Permanent Peoples' Tribunal (PPT) on industrial and environmental hazards and human rights. The PPT was convened at the Yale Law School in the U.S. and then in Bangkok and Bhopal from 1990 to 1992. This Tribunal had no difficulty in coming to a positive conclusion about Union Carbide's guilt and continuing liability.

Usha Ramanathan, an international human rights lawyer, examines initiatives to cover communities from industrial risk in the aftermath of the disaster, while H. Rajan Sharma, who represents Bhopal survivors in the U.S. courts currently, brings the legal history related to the disaster, and its implications, up to date. The chapter concludes with Amnesty International's 2004 segment on the fallout of the compensation procedures.

Batul Bi's compensation nightmare (2004)

The story of Batul Bi, a gas widow with chronic health problems of her own, who has tried for years to get compensation for her husband's death.

Batul Bi, nearly 70, is a resident of Ahata Sikander Kali. Her husband, Taj Mohammad, fell ill after the gas leak and was treated at two private clinics in Bhopal and one in Delhi. He died in September 1989. Batul Bi filed a claim for the death of her husband.

After five years her claim was upheld on 19 June 1995 by a Claim Court of the Deputy Welfare Commissioner. She was granted the minimum compensation of Rs.100,000. However, the Upper Claim Court of the Welfare Commissioner decided, without saying why, to review the case. On 30 August 1996, more than a year later, the Welfare Commissioner set aside the previous decision. The Commissioner's order, about a page long, acknowledged that Taj Mohammad suffered from chronic bronchitis and that his urine thiocyanate test was abnormal. It noted that Taj Mohammad died a day after he was admitted to hospital due to pus formation in his right shoulder, which the Commissioner stated "had nothing to do with exposure to toxic gas," without giving any reason for this explanation. The Commissioner noted that there were no records of the private treatment Taj Mohammad received in Delhi or Bhopal, and concluded: "For the above mentioned reasons Taj Mohammad's death bears no relationship to the toxic gas exposure."

The order downgraded the claim from death to personal injury, ruling that Taj Mohammad should be compensated for his chronic bronchitis, and awarded Rs.35,000. Batul Bi's lawyer-broker then forced her to pay him Rs.32,000 for his efforts. That left her with Rs.3,000. "I spent more than that on my travel, preparing papers and other things. I was left with nothing, except the money that I spent," recalled Batul Bi, almost in tears. Batul Bi filed her own claim for personal injury in early 1988. She has a copy of her registration reference, a copy of the Tata Institute survey that proves she was a resident that night in an affected area, and she is sick. To date, some 16 years later, she has not even received a notification of the hearing of her claim; despite innumerable trips to various offices. The only reply she has received is that her file cannot be found.

Source: Amnesty International report, *Clouds of Injustice: Bhopal, 20 years on*, 2004.

Moral orientations to suffering: Legitimation, power and healing (1995)

Veena Das of Delhi University probes the interface between physical and mental suffering and the behavior of the judiciary and the government around the Indian government's major legislative response to the disaster, namely the Bhopal Gas Leak (Processing of Claims) Act.

Within one month of this tragedy, suits were filed by many American lawyers in courts in America on behalf of several victims. Within a week of the tragedy, several American lawyers had flown into Bhopal and obtained powers of attorney from victims to pursue cases against the Union Carbide Corporation and its Indian subsidiary. These lawyers were described by many as "ambulance chasers" since their fees were said to be based on a contingency sharing of the obtained damages. In any case, in view of the difficulties of pursuing the case by such a large number of poor victims against one of the most powerful multinationals, the Government of India passed the Bhopal Gas Leak Disaster (Processing of Claims) Act in March 1985. With this act, the government of India, in accordance of its *parens patriae* function, took upon itself the responsibility of the conduct of the case and welfare of the victims.

The legal and litigational issues arising out of this case are important and will surely become part of the ongoing debates on toxic torts (Elliot 1988; Rosenberg 1984; Royce and Callahan 1988). Here I shall only describe those events that are necessary for the understanding of one judicial text, *viz.* the judgment of the Supreme Court of India, delivered on December 22, 1989, on the constitutional validity of the Bhopal Act. This is a text in which we can see clearly the professional transformation of the suffering of the victims of Bhopal and the window it opens towards the understanding of how suffering has become a means for the legitimation of the state in India today.

The major events in the litigational history of this episode that need to be kept in mind are the following:

The Bhopal Act was passed by the government of India in March 1985, consolidating all claims arising out of the Bhopal disaster and making the government the only competent authority on the basis of the *parens patriae* function of the state to represent the victims.

In the initial pursuit of the case in 1985 in a New York District Court, the government of India represented that its judicial system was not competent to deal with the complex legal issues arising out of this disaster. The government of India's strategy was to either force Union Carbide to submit to U.S. laws on environmental protection, safety regulations, and liability of hazardous industry, and thus secure compensation for the victims in accordance with law and standards of compensation in the United States, which Union Carbide seemed anxious to avoid; or to ensure that Union Carbide was brought under the jurisdiction of the Indian courts. This was important since the strategy of Union Carbide was to deny that any entity other than its Indian subsidiary, with far lower assets, was liable. Judge Keenan, in whose court this was adjudicated declared faith in the Indian judiciary and bound Union Carbide to the jurisdiction of Indian courts (Baxi and Paul 1985; Baxi 1990).

Hearing of the case began in 1986 in the district court of Bhopal. An application of interim relief was filled on behalf of several victim organizations in 1987, and the judge ruled that Union Carbide should pay Rs.350 *crores* as interim relief, which would be adjusted against final settlement. Union Carbide appealed against this in the High Court, where the quantum of interim relief was reduced to Rs.250 *crores*.

Union Carbide appealed to the Supreme Court against the Madhya Pradesh High Court order. It was while hearing the petition for interim relief that the Supreme Court, to the surprise and dismay of victims, made a judicial order on 15 February 1988, asking the government of India and Union Carbide to settle at $470 million. Part of the settlement deal was immunity granted to Union Carbide and its subsidiaries against any criminal or civil liability arising from the Bhopal disaster at present and in the future.

There were widespread protests against the settlement, although it received the support of some important jurists. It was widely believed that the courts had been pressured or influenced by the Congress government, which was then in power to issue the order and that the government had made a private deal with Union Carbide. Regardless of the truth of these allegations, the order was challenged on several legal grounds. One of these was a challenge to the constitutional validity of the Bhopal Act, since it took away from victims the right to be heard.

The court gave its judgment on the Bhopal Act, holding it to be constitutionally valid, on 22 December 1989. The court was not required to pronounce on the settlement since that was to be heard by another bench; nevertheless, the court made several pronouncements defending the settlement. This particular text is what we shall analyze in detail in this section. To the extent that a judgment creates a master discourse in which the various voices are appropriated in a kind of monologic structure, it is important to see how these voices represent the suffering of the victims to which references abound.

Of the several issues that this case raised, we shall take up only one. How did the government use its *parens patriae* functions, and what did it understand by the act of being surrogate victim as implied by this function? In normal circumstances, *parens patriae* refers to the inherent power and authority of a legislature to provide protection to the person and property of persons *non sui juris*, such as minors, insane and incompetent persons. The victims of the Bhopal disaster had been assimilated to the category of *non sui juris*, even when in the normal course of events they would not be considered judicially incompetent because it was evident that they did not have the resources to pursue the case themselves. However, the Act was challenged on the grounds that by its enactment, the government had constituted itself as surrogate of the victims and taken away their right to be heard. Certain sections of the Act, it was argued, amounted to a naked usurpation of power. The implication of this argument and the timing of the challenge was that having usurped power in this manner, the government had used it to compromise the rights of victims by unilaterally arriving at a settlement and granting immunity to Union Carbide against the expressed wishes of the victims. Thus guardianship in this case amounted to a pretext through which the issue was sought to be resolved against the interests of the victims.

The counter argument to this plea was presented by the then Attorney General, who had represented the government in February 1988 and defended the settlement with Union Carbide in court and outside. In defense of the Bhopal Act he argued that the disaster had been treated as a national calamity by the government of India. The government had a right, and indeed a duty to take care of its citizens in the exercise of its *parens patriae* jurisdiction or in a principle analogous to it. He reminded the court that they were not dealing with one or two cases, but with a large class of victims who, because they were poor and disabled, could not pursue the case against a powerful multinational. In the course of the arguments he maintained that:

Rights are indispensably valuable possessions, but the right is something an individual can stand on, something which must be demanded or insisted upon without embarrassment or shame.

When rights are curtailed, permissibility of such a measure can be examined only upon the strength, urgency, and preeminence of rights and the largest good of the largest number sought to be served by curtailment.

If the contentions of the petitioners are entertained...rights may be theoretically upheld but ends of justice would be sacrificed...the consent of victims should be based upon information and comprehension of collective welfare and individual good.

Here I must make a digression and point out the issues arising out of the necessity for the government to act as surrogate to people who because of disease or handicap (e.g. comatose patients, patients with serious mental trauma) are unable to make their own decisions have been widely debated in recent years. It is generally agreed that the surrogate has to apply a "best interest" standard in making decisions on behalf of people declared to be judicially incompetent (President's Commission Report 1983). However, the interpretation of "best interest" standard has varied. According to one interpretation, the best interest standard requires "a surrogate to do what, from an objective standpoint, appears to promote a patient's good without reference to the patient's actual supposed preferences." Such a view has been challenged on the ground that a surrogate is acting in the best interest of the patient only if he tries to replicate the decisions that a patient would have made if he was capable of doing so. Evocation of objectively determined interests as against subjectively expressed preferences can simply mask power in the name of doing good.

The same question now presented itself before the Supreme Court. In its earlier judicial order, the Supreme Court had claimed that it had acted in the best interest of the victims. But the victims were contesting this. How could the court defend the actions of the government and its own actions in promoting the settlement, when victims themselves were protesting against it? It is in this context that the "suffering" and the "agony" of victims seems to have found its *raison d'etre* of allowing the judiciary to create a verbal discourse which would legitimize the position of the government as the guardian of the people, and the judiciary as the protector of the rule of law.

In their judgment, the judges upheld the constitutional validity of the Bhopal Act. In a stirring statement they declared, "Our Constitution makes it imperative for the State to secure to all its citizens the rights guaranteed by the Constitution and when the citizens are not in a position to assert and secure these rights, the State must come into the picture and protect and fight for the rights of the citizens." This is a principle that expands greatly the power of the state to act as guardian and deprive people of their normal constitutional rights including the right to petition a court of law. Aware of this contradiction the judges further stated, it is necessary for the state to ensure the fundamental rights in conjunction with the Directive Principals of state policy to effectively discharge its obligations and for this purpose, if necessary, to deprive some rights and privileges of the individual victims in order to protect these rights better and to secure them further." But in what way had the meager settlement served to secure the rights of the victims or led to "greater good of greater numbers" – the utilitarian principle formulated by the attorney general and the judges at several times in the course of the argument?

The judgment stated that the Court had, in its earlier judicial order which instructed both parties to settle on the conditions described earlier in the paper, acted in the interest of the victims. The basic consideration, the court recorded, motivating the conclusion of the settlement was the compelling need for urgent relief, and the Court set out the law's delays duly considering that there was a compelling duty both judicial and humane, "to secure immediate relief for the victims." Here is the first invocation of the suffering of the victims; since there was a compelling need to provide immediate relief, the Court had been moved by humane considerations to instruct parties to settle. The Court simply neglected to mention that the urgent need for immediate relief was precisely what the victims had asked for in their petition for interim relief, which had been upheld by lower courts! Further, immediate need for relief was simply stated as the motivation behind settlement, but there was nothing in the judicial order that would instruct government to provide this immediate relief – neither the procedure for dispersal of money nor any timetable having been formulated by the Court. One would not, therefore, be incorrect in concluding that the suffering of the victims was more a verbal ploy than a condition that caused serious concern to the judiciary or the government.

On the question as to why the Court had not found it fit to invite the opinion of victims themselves, at least through some representative organizations, about the proposed settlement before passing the judicial order, the judges noted that this had been because of a certain degree of uneasiness and skepticism expressed by the learned counsel of both parties "at the prospects of success in view of their past experience of such negotiations when, as they stated, there had been uninformed and even irresponsible criticism of the attempts to settlement." Thus the first imperative to arrive at a meager settlement was the "suffering" of the victims. The second imperative to do this with complete secrecy and present victims with a *fait accompli* was the irresponsibility and inability on the part of victims to understand the issues.

Thus the situation was as follows: A multinational corporation was engaged in the production and storage of an extremely hazardous industrial chemical for which it had been given the license to operate by the Indian government. Despite the known hazards of industrial isocyanates and diisocyanates, neither the multinational corporation or its Indian subsidiary, or the government of India had considered it important or necessary to inquire into the nature of hazard to the people posed by the activities entailed in its manufacture and storage at the various times between the setting up of the factory and the spillage of the gas. The people of Bhopal, especially those staying around the factory, had not been warned of the dangers posed to them by these industrial activities, nor had any regulations been made and implemented about the placement of such factories. The result of all these activities geared towards the "development" and "industrialization" of India was that more than 300,000 people were suddenly one night blighted by crippling disease and more than 2,500 died horrible deaths. Yet the people to be declared incompetent and irresponsible were neither the multinational nor the government, but ironically the sufferers themselves.

How had the Court arrived at the particular sum of 470 million dollars as adequate compensation for the 300,000 victims already known to have been affected? Were the victims right in alleging that there was no evidence of the application of judicial mind behind this order? The fact was that medical examination of only half the claimants out of the 600,000 claimants had been completed at the time that the court-ordered settlement, through procedures that were highly questionable. Similarly, the medical folders that recorded the basis on which a victim's status was determined for purposes of compensation were not made available to them or their representatives on grounds of medical secrecy. Thus bureaucratic decisions could not be contested by victims.

Incidentally, it should be noted here that since the 22 December judgment, the victim groups have been able to secure some primary documentation on the procedures through which extent of victimization was estimated. It has been found that even the figures produced in the Supreme Court during the hearing by Madhya Pradesh Council were arrived at after the settlement, and therefore could not have been the basis for the settlement. Further, the procedure by which the status of a claim was determined was based on a kind of scoring system by which each symptom was given a score and finally the scores were aggregated. Despite the fact that the medical journals have constantly used the rhetoric of MIC being the most lethal gas known to man, the scoring system used the analogy of organ by organ injury rather than systemic damage as a result of toxic insult. To compound it all, patients who could not produce documentary evidence in the form of records of hospital admission or proof of having been treated in the first few days of the gas disaster were declared to be "uninjured" regardless of the state of their health at the time of examination. This disregard of all human and medical ethics is appalling, for anyone familiar with mass disaster would know that the immediate task at the time of disaster is to reach help and not to keep meticulous individual records. All this has been justified on the grounds of bureaucratic and legal necessity. The macabre quality of this discourse can be seen in the fact that Union Carbide has repeatedly tried to present itself as the real "friend" of the victims whose offer of "the best medical treatment for victims" was rejected by India. In making such arguments Union Carbide lawyers simply fail to mention the fact that at the same time that these offers were being made, the medical scientists flown into Bhopal by Union Carbide were declaring that MIC was not a dangerous gas at all and that if victims had died, it was their own fault. To take just one example of this attitude, consider the statement of Dr. Peter Halberg, one of the three doctors sent by UCC as part of its 'relief efforts' to Bhopal. According to the learned doctor, "MIC created a heavy cloud, which settled very close to the earth, killing children because of their immature lungs, the elderly because of their diminished lung capacity, those who ran because their lungs expanded too rapidly." The transformation of healers into merchants and touts of death unfortunately has always been a possibility inscribed into the medical discourse, as the labors of Robert Lifton have proved. How was this potential to be resisted?

Here we can see that disabled though the victims were they did not let this particular construction of their reality go unchallenged. In response to the question as to how the Court at the time of settlement had arrived at the judgment that the sum was reasonable, it had been stated in the clarificatory orders that the reasonableness of the sum was based not only upon independent quantification, "but the idea of reasonableness for the present purpose was necessarily a broad and general estimate in the context of a settlement of the dispute and not on the basis of an accurate assessment of adjudication." In other words, victims were irresponsible and uninformed when they inquired about the basis on which the Court had arrived at the estimate of the victimization or the principles by which compensation was being determined, but the Court itself had only been motivated by humane considerations when it refused to divulge any of the medical information about the nature and extent of damage caused by MIC.

In its final transformation then, the suffering of the victims was considered sufficient reason to justify the settlement and to uphold the usurpation of power by the government on the principle that the Bhopal Act could be upheld if it had proved to be in the interests of the victims (the proof of the pudding is in the eating – as the learned judges stated). Therefore, although the judges conceded that justice did not *appear* to have been done (since the victims were not given a hearing), they also argued that justice was, nevertheless, *done* on the principle that a small harm can be tolerated for a larger good. The judges were concerned to signal their own humanity and to protect the legitimacy of the judicial institution of which they were a part. The "suffering" of the victims was a useful narrative device which could be evoked to explain why victims had not been consulted; why their protests over the settlement could be redefined as actions of irresponsible and ignorant people; and above all, why the judges had not felt obliged to ask the government and its medical establishment to place for public scrutiny what it had accomplished by way of providing relief and help to the victims, and in the process, what was the knowledge that had been generated on the impact of the deadly isocyanate on the health of people; and finally, that they had completely failed to fix responsibility for the accident and thus had converted the issue of multinational liability into that of multinational charity.

In the course of the judgment; it was evident that there was no lack of concern for the impact of hazardous industry on *society in general*; it was only the interests of these particular victims that could not be fully protected. Thus there were acute reflections on the dangers arising from hazardous technology. The judges stated that there were vital juristic principles that touched upon problems emerging "from the pursuit of such dangerous technologies for economic gains by multinationals." The judges also noted the need to evolve a national policy to protect national interests from such ultra-hazardous pursuits of economic gain. However, the sufferings of the people dearly came in the way of evolving these juristic principles. To quote the judgment on this issue:

In the present case, the compulsions of the need for immediate relief to tens of thousands of suffering victims could not wait till these questions, vital though these may be were resolved in due course of judicial proceedings; and the tremendous suffering of thousands of persons compelled this Court to move into the direction of immediate relief, which, this Court thought, should not be subordinated to the uncertain principles of law...

In all these discussions at the judicial and bureaucratic levels, "victim" finally became a totally abstract category. The judges did not seem to have any clear idea about how many victims there were or the havoc caused by MIC poisoning on their bodies and minds. In a moving speech protesting the settlement, one illiterate woman stated: "We only ask the judges for one thing – please come here and count us." In the judicial discourse, however, every reference to victims and their suffering only served to reify suffering, and to dissolve the real victims so that they could be reconstituted again into nothing more than verbal objects.

Source: Das, Veena, Moral Orientations to Suffering: Legitimation, Power and Healing in ed. L.Chen, et al., *Health and Social Change in International Perspective*, p. 139. Boston: Harvard Series on Population and International Health. (Footnotes omitted.)

Permanent Peoples' Tribunal: Conclusions and judgment in Bhopal (1992)

Sessions of the Permanent Peoples' Tribunal on Industrial and Environmental Hazards and Human Rights took place at the Yale Law School in the U.S. and then in Bangkok, Bhopal and London from 1990 to 1994. In this excerpt the tribunal issues a decision on the Bhopal disaster.

Having received both oral and written testimony from victims, workers, experts and others at all three sessions of the Tribunal on industrial and environmental hazards and human rights (Yale, Bangkok, and Bhopal) the Tribunal finds that:

1. The fundamental human rights of victims of the world's worst industrial disaster, including Articles 1, 7, 10, and 16 of the Universal Declaration of the Rights of Peoples (Algiers, July 1976), Articles 1, 2, 3, 5, 7, 8, 12, 17, 19, 23 and 25 of the Universal Declaration of Human Rights

(New York, December 1948), and Articles 14, 19, 21, 38, 39, 41, 42 and 43A of the Indian Constitution have been grossly violated.

2. Union Carbide Corporation, its subsidiary, Union Carbide (India) Ltd., and key officials of both are clearly guilty of having caused the world's worst industrial disaster through the design and operation of the Carbide factory in Bhopal, regardless of the extent to which others, including local, state, and national government, may have contributed to the disaster. This finding is based on Carbide's own documents, the evidence generated through the discovery process in the Federal District Court in New York, and oral and written testimony presented at three sessions of this Tribunal.

3. The Government of India and the Government of Madhya Pradesh are also partially guilty of violating the rights of the victims, not only under international human rights law, but also under the Indian Constitution. The catalogue of wrongs inflicted on the victims by these governments is long, beginning with location of the extremely hazardous MIC unit in a populated area in contravention of the government's own rules and the urban plan for Bhopal. The catalogue continues with such wrongs as the deliberate mis-categorization of victims in terms of injury or disablement, the refusal to register a large number of claims, including many children, and the inexcusable delay in processing claims.

4. Existing mechanisms to deal with the consequences of such catastrophes, and prevent them happening in the first place have failed miserably thus far in Bhopal. Equally conspicuous for their failure are the two prevailing modes for dealing with threats to the environment and human safety namely, industry self-regulation and government policing.

5. Also conspicuous in their failure to help ameliorate the distress of the victims have been, with some notable exceptions, the legal and medical professions. Indeed, on some occasions these professions have aided and abetted the re-victimisation of the victims.

6. Under no circumstances can the killing and injury of so many innocent people be considered an acceptable cost of development. Such a rationalization reflects a twisted and pathological approach to social and economic changes.

Judgment

Based on these findings, the Tribunal regrets and deplores:

The manner in which the due process of adjunction was circumvented by the premature and unwarranted announcement of the February 1989 order imposing an unjust settlement on victims without consulting them.

This order is all the more deplorable because it allowed UCC to escape without facing legal or financial responsibility, because the amount of the settlement is too little to meet existing needs, and because the unmet burden will fall on the Indian taxpayer rather than the guilty party.

The prevention, through this settlement order, of completion of hearings on the merits in the case.

The justification of the settlement by way of a manipulated medical categorization which violated the right to justice of victims, and virtually accomplished denial of expeditious and equitable realization of damage and compensation.

The failure of the Indian Supreme Court to seize the opportunity to keep Union Carbide in its continuing jurisdiction for a definite period of time to ensure that its duties of cooperation in recognizing rights of victims and meeting their legitimate claims arising from the latent, unfolding, and long term effects of MIC would be fulfilled.

The lack of proper health care, which has been denied to most of the victim community.

The denial of the basic dignity of victims, who have been subject to highly arbitrary bureaucratic processes, such as the administration of interim relief, review of victim claims, and unkept promises of medical treatment. Victims of industrial and environmental hazards have a nonnegotiable right to dignity which has not been respected in Bhopal.

Policies of urban destruction which revictimize the victims by destroying their impoverished access to the right to shelter and callous diversion of funds intended for the rehabilitation of gas victims for urban development and beautification unrelated to their real needs.

The failure of government to provide sustained and meaningful opportunities for the vocational rehabilitation of the victims so that they would no longer be dependant on government doles.

The Tribunal appreciates the following developments as historically appropriate responses that set precedents for comparable action in other cases involving violations of human rights from industrial and environmental hazards:

The provision, following an exceedingly long and concerted campaign of protest and lobbying by the victims, of *interim relief* aimed at helping victims resist the imposition of unfair or inadequate settlements, even though the relief started more than four years after the disaster and its administration has been marred by corruption.

The rigorous enunciation and pursuit of the legal doctrine of *absolute multinational enterprise liability* by the government of India which we urge be universalized through national legislation and international agreement.

The lodging of *criminal charges* against Union Carbide Corporation, its Indian subsidiary, the key officials of both companies.

The Tribunal recommends, with the utmost concern for immediate and effective action, that the following steps be taken without delay:

Development and implementation by the governments of India and Madhya Pradesh of a comprehensive plan of action for economic and medical rehabilitation of victims, formulated with the participation of victim groups and accountable to both, the local and international communities.

Assurance by the national and state governments of continued access to shelter in accordance with international standards set forth in international human rights instruments, such as the Universal Declaration of Human Rights, the Covenant of Economic, Social, and Cultural Rights, and the Declaration of Right to Shelter.

Development of clear biomedical criteria for compensation of victims, with satisfactory provision of those victims unable to document the connection between their current disability, eight years after the disaster, and exposure to MIC and other gases from the Carbide pesticide factory. The distribution of compensation must not only avoid false claims, but also must avoid unjust exclusion of true claims. The government must expect that a large number of victims will have legitimate claims and yet be unable to document their residential and medical histories. An inability to provide documentation cannot be an excuse for re-victimization.

Voluntary organizations and activists should consider how they can help local communities organize by neighborhoods, identify their victims, and assist them in documenting their expenses during and since the disaster. Action should be taken to restore 7 out of 15 categories of compensable claims as set forth in the Bhopal Gas Leak Disaster (Registration and Processing of Claims) Scheme of 1985 and as directed by the Government of India Ministry of Chemicals and Fertilizers in its letter to the Madhya Pradesh Welfare Commissioner for Bhopal Gas Victims.

Provision of adequate, readily available, and continuing health care and monitoring.

Creation of an independent international Medical Commission to review available data on the health conditions of the victims, to identify gaps in the data with recommendations of how, if at all, they can be filled, and to propose realistic criteria for determining how the victims, on the basis of available and readily obtainable data can be classified. This Commission might also be designated as a body to which individual victims could appeal their medical categorization if the government persists with its present effort to determine medical conditions on the basis of existing health records and superficial medical examinations.

Establishment of an international panel of medical and economic specialists in compensation to determine fair compensation levels for different categories of disabled victims and for relatives of the deceased.

An independent investigation into the harassment of women victims, including police violence. The investigation should also examine attempts to interfere with their efforts to organize for mutual protection and advocacy of their rights. The investigation should consider violations of rights as set forth in the Indian Constitution and international law including the Convention on the Elimination of All Forms of Discrimination Against Women.

A comparative study by an independent body of competent persons of the safety systems and procedures in the Carbide pesticide factories in Institute, West Virginia, and Bhopal to determine to what extent the Bhopal disaster can be attributed to double standards between similar plants in industrialized and Third World countries.

Prompt issue of an extradition order for former Union Carbide Chairman Warren Anderson under the Indo-U.S. extradition treaty and vigorous and timely prosecution of Warren Anderson and the Union Carbide Corporation for culpable homicide as stipulated under the Indian Penal Code and in fulfillment of the Supreme Court decision of October 1991 restoring the original criminal charges.

Source: The Permanent People's Tribunal, Bhopal India.

Industrial risk in Indian law (2004)

According to Usha Ramanathan, an international human rights lawyer, industrial risk was a dormant concern in Indian law until the Bhopal Gas Disaster. The following excerpt from her essay, "Communities at Risk", traces the impact of Bhopal and the steps that have been taken through legislation and litigation to protect communities at risk.

Evidence of emerging concerns regarding safety and harm is discerned in the enactment of laws including the Air (Prevention and Control of Pollution) Act, 1981 and the Water (Prevention and Control of Pollution) Act, 1974 which grew out of an increasing awareness about the need to control pollution occasioned by industrialization and urbanization. These hold the first glimmerings of understanding that what happens on the premises of a factory could affect persons and communities well beyond its boundaries. Yet it was not till the Bhopal Gas Disaster that the designation of a community at risk was dramatically altered.

In the intervening night of 2–3 December 1984, methyl isocyanate (MIC) leaked out of the Union Carbide plant in Bhopal. It left over 3,000 dead in its wake, and over 5,00,000 people affected by the gas in varying shades of severity. More than 15,000 people have died since that day of gas related illnesses. It has been variously called "a grim tragedy", "one of the most tragic individual disasters in the recorded history of mankind" and "second in magnitude and disastrous effects only to the havoc brought by atomic explosions in Hiroshima and Nagasaki" by judges deciding the Bhopal matters in the Supreme Court. The Bhopal Gas Disaster, demonstrating the potency of industry to cause mass death, injury, illness and loss well beyond the perimeters within which its activities are located, altered industrial risk irremediably, and expanded it into industrial hazard. The constituency of realized risk stood definitively changed. The induction of the "general public in the vicinity" of the factory into the law, as potential victims of industrial hazard, is a manifestation of this change.

The Emergence of a Community at Risk

Till the Bhopal Gas Disaster happened, industrial risk was linked with occupational health and safety. The constituency of industrial safety law was the workforce, and the site of industrial risk was the premises where industrial activity was carried on.

Immediately, and directly, the victims of the disaster emerged as an identifiable constituency that the law had to acknowledge. The victims bore these characteristics:

- That their residence was around the factory. Of the 56 wards into which the city of Bhopal is divided, the residents of 36 wards were affected by the disaster in different measures of severity.
- Their habitation was in the nature of an industrial shanty town. The houses were packed close together, and the residents belonged to the working classes.
- While there were those among the victims who lived in the area and worked in the Union Carbide plant, they largely comprised persons who had no connection with the industry beyond being residents in the vicinity.

The litigation around compensation, criminal prosecution of the company officials, medical relief and treatment, and residual and spreading contamination in and around the factory, followed a tortuous course over the succeeding years; in fact, much of it remains to be concluded in 2004.

In February 1989, the Union of India, representing all the victims of the disaster settled the claims for compensation for $475 million. The categorization of claims arising out of the disaster began after challenges to the settlement by victims' groups and public interest petitioners had been decided by the Supreme Court in 1991. The exercise continues with over 16,000 claims still to be decided. On July 19, 2004, the Supreme Court ordered that the monies which remained under control of the Union of India belonged with the victims, and that it be disbursed *pro rata* among them; this process is yet to begin. The criminal trial of the offending company and its officials and of the officers of its Indian subsidiary is still under prosecution in the court of first instance, with the Union Carbide Corporation (UCC) and Warren Anderson, the CEO at the time of the disaster, being declared absconders. After years of agitating for treatment and monitoring of the health of the victims, the court has conceded a demand for setting up an independent medical commission. The soil and ground water contamination in and around the factory, not all of it caused by the disaster but aggravated by the neglect and abandonment of the site following the disaster, was detected and reported upon by Greenpeace in 1999, and a court in the U.S. in March 2004 declared its willingness to order a clean-up, and the government of India agreed to let Union Carbide and Dow Chemicals (who have taken over UCC) comply with the U.S. court's direction. This process is yet to begin.

In the meantime, a day and a year after the Bhopal Gas Disaster, an industrial accident in Delhi provided the impetus for reconsidering the context of industrial risk and disaster. On 4 December 1985, oleum gas leaked into the atmosphere in Delhi, spreading from its source in the factory of Shriram Foods and Fertilisers, moving through populous zones, affecting large numbers of the public that it encountered on its way, reportedly resulting in a death of an advocate in the Tis Hazari courts caused by inhalation of the gas. Even as the enormity of the Bhopal Gas Disaster paralysed parts of the apparatus – and even as disinformation, lack of information and unpreparedness aggravated the direct damage to those affected by that disaster – the oleum gas leak, with the relatively limited extent of injury and loss, allowed an immediacy to enter the discourse. Where Bhopal showed up the vulnerability of the industrial shanty town, the oleum gas leak raised the spectre of the denizens of Delhi living under perpetual threat posed by hazardous industry. The populations at risk were identified by the experience of realized risk in one instance, and the demonstration of potential risk in the other.

The oleum gas leak provoked a range of responses including:

- legislative changes, particularly to the Factories Act 1948 in 1987,
- essaying a policy of deterrence through an enunciation of enterprise liability, and
- the induction, into policy, of relocation of hazardous industry away from concentration of populations.

Amending the Factories Act 1948 in 1987

There are three aspects of the amendment made to the Factories Act 1948 in 1987, incorporating the experience of the Bhopal Gas Disaster and the oleum gas leak, which are directly related to a recognition, and anticipation, of a community at risk. Abandoning the presumption that industrial risk is related only to the workforce and is restricted to the premises of a factory, parliament acknowledged the risk posed to "the general public in the vicinity" by a factory "involving a hazardous process." The distancing of the population at risk from the factory, especially of those factories already established, being an improbable solution to the question of risk, the amended law instead requires "compulsory disclosure of information" "regarding dangers, including health hazards and the measures to overcome such hazards arising from the exposure to or handling of the materials or substances in the manufacture, transportation, storage and other processes." With the introduction of this provision into the law, *the recognition of a community at risk was expanded beyond the workforce to include persons in the vicinity of the risk.* The factory is still the site from where the risk emanates, but the effect of realized risk was acknowledged as reaching beyond the premises of the factory.

The second aspect was concerned with prospective location of industrial risk. There is an inevitable constriction of choice where a factory has already been established and is in operation, and where a population has settled around it. Against the background of the Bhopal Gas Disaster and the oleum gas leak, the realization of risk, even disaster, has necessarily become a part of law and policy. A Site Appraisal Committee has, therefore, been introduced by the law "for purposes of advising (the government) to consider applications for grant of permission for the *initial location* of a factory involving a hazardous process or for the *expansion* of any such factory." Among its members are to be an expert in the field of occupational health, a representative of the Town Planning Department of the state government, water and air pollution control authorities, a representative of the meteorological department of the government of India, a representative of the local authority and the chief inspector of factories.

Factories in operation, and those being established, are required to draw up "on-site emergency plan and detailed disaster control measures." And "workers employed (in the factory) and the general public living in the vicinity of the factories" are to be informed of "the safety measures required to be taken in the event of an accident taking place." This is the third aspect of risk recognition that has been enacted into law.

Compensation, Liability and Deterrence

There are three approaches that law could offer for reducing risk and enhancing safety. The first is *preventing risk realization*. So, authorizing workers to alert the inspector of factories, and the management, to lapses of safety; and monitoring and regulating the safety features in the running of a factory, are moves towards preventing accidents and disasters.

The second is the threat of *criminal prosecution*, both under the Penal Code 1860, which deals with crimes, and as "absolute offences", where breach of a provision in the law becomes punishable even where no intention to commit an offence need be proved before punishment is visited upon the person found to have been responsible – by neglect or by commission. Breaches of provisions in the Factories Act 1948 fall within the law's understanding of absolute offences.

The third is *deterrence through a deep pocket approach to compensation*. After the oleum gas leak, for instance, a director of the offending company was called upon to undertake to pay compensation from his own resources – not to be reimbursed by the company – in the event of a further leak, before the factory was permitted to restart its operations. The Supreme Court also brought on board a notion of "absolute liability" and "enterprise liability", which would strengthen the enterprise's interest in ensuring safety.

Industry Relocation

The most dramatic response to industrial risk has been relocation. The ascension of risk to a hazard, and the experience of that hazard in Bhopal, has altered the perception of acceptable risk. The growth of settlements around factories has been both inevitable, and by policy illustrates the emergence of a community of those officially housed by the company in the vicinity of the factory, and employees in jhuggies and "outsiders" allowed to settle around the factory premises. Bhopal had demonstrated the vulnerability of the population around the factory to realized risk; and the oleum gas leak had reinforced fears that whole populations of cities could be at enormous risk. The exiling of risk and hazard was one possible response; the minimizing of risk was another.

Conclusion

There is an incoherence in the development of law and policy on industrial risk. The anxiety about risk and hazard exists, but legal imagination has not been able to cope with the consequences of either leaving risk where it is, or exiling it. The courts do not possess the equipment needed to work out the reorganization of spaces to minimize, or outlaw, risk. Yet, when the question of risk and hazard is taken to the court, the court cannot turn away. It has sometimes refused to be definitive, and sometimes shown a tolerance of risk, asking of persons resident around risk to become superior risk bearers. The unease of the law persists.

Source: Usha Ramanathan, Communities at Risk: Industrial Risk in Indian Law, *Economic & Political Weekly*, October 9, 2004, [footnotes omitted.]

Catastrophe and the dilemma of law (2004)

H. Rajan Sharma, a lawyer who is presently representing residents of Bhopal in a Federal class action against Union Carbide in U.S. courts, brings the legal story up to date, outlines his case, and looks at what Bhopal means for the future of Human Rights.

> The juridical field is the site of a competition for monopoly of the right to determine the law. Within this field there occurs a confrontation among actors possessing a technical competence which is inevitably social and which consists essentially in the socially recognized capacity to interpret a corpus of texts sanctifying a correct or legitimized vision of the social world. Such a process is ideal for constantly increasing the separation between judgments based on the law and naive intuitions of fairness.
>
> – *Pierre Bourdieu*[1]

Around midnight on 2–3 December 1984, a pesticide plant leaked some twenty seven tons of methyl isocyanate, a highly toxic chemical gas, into the air over the sleeping city of Bhopal. The results took the form of an enormous tide of death that welled up in the city's hospitals and morgues. Official estimates, probably understated, put the toll at a staggering 2,000 dead but reliable unofficial estimates suggest a soul-numbing figure of nearly 6,000 or more dead in the 48-hour aftermath of the disaster. The cause of death was described in most cases as pulmonary edema, a polite medical euphemism for an excruciating manner of death by slow drowning in one's own bodily fluids. According to the Indian Council for Medical Research, more than 250,000 people continue to suffer from permanent disabilities and chronic ailments as the result of exposure to the poisonous gases on that night. This December marks the twentieth anniversary of this unparalleled disaster, perhaps the single worst industrial catastrophe ever to befall a civilian population. By some accounts, at least 20,000 persons have died over the past two decades. The International Commission for Medical Research on Bhopal has concluded that, due to chromosomal and genetic damage among the victims, the wake of this unprecedented catastrophe will continue to ripple through the next three to four generations in Bhopal in the form of birth defects.

The word "tragedy" has shown a talismanic insistence in appearing and reappearing in the context of this incident. Perhaps it helps us soothe the conscience by lending an air of the inevitable or unavoidable to these events or maybe it helps us invoke that Faustian mythology so typical of modern society in which nameless others are sacrificed at the hallowed altar of progress for our technological hubris. But this anniversary should not just memorialize the tragedy of Bhopal. It should be an occasion to recall the travesty of what the victims have been made to endure over the past twenty years.

Despite all the fine sentiments and noble ambitions expressed in what the Indian Supreme Court, in this case, referred to as the "uncertain promise of law," the fact of the matter is that the law, in all its abstract majesty, has utterly failed to provide the victims of the world's worst industrial disaster with so much as a semblance of justice over the past two decades. Instead, the law has been the principal author of a kind of Kafkaesque parody of justice that has played itself out in the courts of the United States and India. The so-called wheels of justice have, in this case, turned only to crush the hopes of the survivors beneath them. The seemingly endless processes of the law have, in fact, perpetuated and compounded an injustice too fearful to contemplate, which has been allowed to stand without redress or remedy for twenty years, seven thousand three hundred days (to be precise), each day a shameful vindication of the maxim that holds that laws are like cobwebs, strong enough to only detain the weak and too weak to constrain the strong. The epitaph has yet to be written on the sordid record of what may justifiably be called the Bhopal travesty.

The plant belonged to Union Carbide India Limited, an affiliate of Union Carbide Corporation, a multinational corporation headquartered in the United States. Union Carbide owned 50.9% of its Indian affiliate at the time of the disaster and was responsible for transferring proprietary technology to the Bhopal plant for manufacturing Sevin, a patented product in which methyl isocyanate was one of the key ingredients. It was determined shortly after the disaster that a "runaway reaction" of a highly volatile chemical, methyl isocyanate or MIC, had taken place in one of the plant's storage tanks. The most likely cause was the introduction of water into the tank. It is undisputed that a routine "water-washing" operation prescribed by the American company's operating manuals were being conducted on the night of the disaster.

Numerous independent investigations have concluded that, while the entry of water into the storage tank may have triggered the runaway reaction, the real causes of the catastrophe can be traced to the decision to store methyl isocyanate in large quantities for long periods of time, the badly flawed design of the plant as well as the near-total absence of safety provisions and emergency-preparedness measures. Needless to say, Union Carbide has strenuously contested this version of events, disclaiming any managerial responsibility for the design, day-to-day operation of the UCIL facility or its safety features and asserting that its relationship with its Indian subsidiary was a hands-off or arm's length relationship. The survivors and their representatives, meanwhile, have maintained that the U.S. company deliberately chose to bequeath to Bhopal an obsolete, dangerous and ill-equipped plant, with grossly inadequate technology, pointing to Carbide's methyl isocyanate facility in Institute, West Virginia as an example of the company's discriminatory imposition of double standards of risk, safety and emergency-preparedness. The Institute plant, they argue, was designed with significantly higher parameters for safety and emergency- preparedness: e.g., computerized warning systems, larger capacity safety devices, and safer processes for storage and containment of methyl isocyanate. For the past two decades, Carbide has insisted that standards of design, technology, safety and emergency-preparedness were either uniform or at least comparable at all of its worldwide operations, including at Bhopal. To date, Union Carbide continues to withhold scientific and medical research on the toxicology of the leaked gases which could assist in the treatment of innumerable victims on the specious grounds that this information constitutes a "trade secret."

In the disaster's aftermath, hundreds of lawsuits were filed in jurisdictions across the U.S. against Union Carbide by American contingency-fee lawyers. These were ultimately consolidated into a single proceeding before Judge John Keenan in the Southern District of New York. Fearing that the victims claims might be exploited by an army of private lawyers, the Indian Parliament enacted the Bhopal Gas Leak Disaster (Processing of Claims) Act, on 25 March 1985. The legislation was based on a doctrine under international law known as *parens patriae* (literally, "parent of the country"), which held that the State was empowered to act the legitimate protector of its citizens and their environment. The Act conferred upon the Indian government the full power and authority to act as the exclusive legal representative of the survivors in all claims for compensation before foreign or domestic courts, subject to its obligation to consult and cooperate with the victims and their representatives in the prosecution of such claims. Framers of the legislation argued that it would enable the Indian government to provide centralized, integrated decision-making and control to prosecute claims on behalf of the mostly destitute victims, bringing all the nation's resources to bear against the multinational might of the corporation.

Based on this legislation, the Indian government filed suit against Union Carbide on 8 April 1985, in the courts of the United States, where, in what can only be described as a profoundly ironic exercise, India argued that the interests of justice required the case to be tried in the United States on the grounds that its own legal system was backward and procedurally outmoded, lacking any class action device or other provision for representative suits, burdened with the legacy of colonialism, and subject to massive delays caused by endemic docket backlogs. The company countered that the case ought to be tried in the courts of India, without burdening American taxpayers, and showered praise upon the legal system of the "world's largest democracy", particularly to the extent it was "based on sound and established principles of Anglo-Saxon law." On 12 May 1986, Judge Keenan conditionally dismissed the consolidated action on the grounds that India was the more appropriate forum for the resolution of this litigation. The decision rested, in part, on the notion that trying the case in the US courts would amount to "yet another instance of imperialism" imposing foreign legal standards upon a developing country with "vastly different values", different levels of "population" and "standards of living." The dismissal was conditioned upon Union Carbide's "consent to submit to the jurisdiction of the courts of India." Meanwhile, criminal proceedings and investigation had already been initiated against Union Carbide and its former director, Warren Anderson, in the Bhopal District Court in 1984 and formal charges of "culpable homicide" and "causing death by use of a dangerous instrumentality" were framed by India's prosecuting agencies on 30 November 1987. The charge of culpable homicide under the Indian Penal Code is equivalent to manslaughter, causing death by reckless indifference.

By the time the case first reached the Supreme Court of India on the issue of whether interim relief assessed against Union Carbide on behalf of the victims was appropriate, litigation had continued in India for more than five years without even reaching the commencement of pretrial discovery. The mostly destitute victims had received nothing in the way of compensation from their erstwhile "parent," the Union of India. On 14 February 1989, Chief Justice Rajinder S. Pathak interrupted the proceedings to announce that he felt, in light of "the enormity of suffering occasioned by the Bhopal gas disaster and the pressing urgency to provide immediate and substantial relief to the victims," that the case was "preeminently fit for overall settlement."[2] The Chief Justice then entered a judicial decree preliminarily recording the terms of this settlement which required Union Carbide to pay $470 million in damages in order to extinguish all civil and criminal liability.

The Indian Supreme Court, however, gave the victims and their able counsel a last opportunity to challenge the terms of the proposed settlement. In its final decision of October 1991, the Indian Supreme Court justified the settlement giving it final approval on the grounds that the victims' fate could not be left to the "uncertain promise of law," but modified its terms and conditions by mandating the prosecution of criminal charges against Union Carbide and its former director, Warren Anderson, which had been pending since 1987.[3] Criminal charges against Union Carbide and Warren Anderson were accordingly renewed in the Bhopal District Court. In March 1992, the Judicial Magistrate issued an arrest warrant for Warren Anderson and gave lawyers for Union Carbide a month in which to appear for trial. Neither of the parties presented themselves in court and Union Carbide's official spokesperson stated that the company would flatly refuse to submit to the jurisdiction of India's criminal courts. Summons served on Union Carbide through the U.S. Department of Justice were ignored. In 1992, the Bhopal District Court proclaimed the company and Mr. Anderson to be "absconders", i.e. fugitives from justice. To date, neither Union Carbide nor Anderson have appeared to face the criminal charges pending against them in India. As recently as 2004, the Indian government submitted an extradition request for Anderson under the Indo-US Treaty of Extradition which was, reportedly, rejected. Progress in the criminal case against Indian officials has been, if anything, equally glacial.

The compensation tribunals, established by the Union of India, the erstwhile "parent" of the disaster victims, did not even begin functioning until 1992. Present-day estimates by non-governmental organizations indicate that over 90% of claimants have received less than $400 from the claims process in India, an amount insufficient to pay for medications over a five year period.

Meanwhile, tests of the water supply of as many as sixteen communities residing near the plant site and surrounding environs have revealed severe environmental contamination of the aquifer in Bhopal resulting from the indiscriminate disposal of toxic chemicals and by-products produced there. Recent soil and water sample tests conducted by an independent British laboratory for Greenpeace and certain victims' organizations indicate massive contamination of soil and drinking water around the facility in Bhopal. To cite one instance, the Greenpeace report states that water samples taken from the Bhopal site contained carbon tetrachloride, a carcinogenic chemical, which exceeded maximum tolerance limits established by the World Health Organization by 1,705 times. Union Carbide claims that it has no further responsibility or liability for the environmental remediation of the plant site because it has sold its shares in its Indian subsidiary and the land was returned to Madhya Pradesh in 1998.

In litigation before the Indian Supreme Court, the Union of India has sought to utilize the interest earned on the settlement fund, over the many years that it remained undistributed, to clean-up and remediate the badly polluted plant site and the groundwater aquifer which provides the drinking water of as many as 20,000 residents of affected communities. The Supreme Court has, mercifully, denied its request and ruled that the victims must receive the remainder of these sums to which they are legally and morally entitled. But lawyers for political parties have filed objections claiming, with truly democratic largesse, that these funds should be allocated to a dozen or so municipal wards in Bhopal where the effects of the disaster were felt principally in the form of a temporary inability to find good maids and kitchen help.

As an attorney who has had the privilege of representing the survivors' cause in the courts of the United States, it pains me to admit that my own efforts have achieved only modest success in turning the tide of this battle.

That litigation, commenced in 1999, consisted primarily of an effort to translate the unresolved criminal liability of Union Carbide into an actionable violation of international human rights law. The complaint also included claims for damages to physical health and property as a result of the contamination of drinking water as well as for injunctive relief in the form of clean-up and remediation by Union Carbide of the severely polluted land for its factory, which still has thousands of metric tons of waste stored above-ground and buried in a landfill on site, and the aquifer in Bhopal. Those efforts were unavailing largely because the American courts gave scant weight to our argument that Union Carbide could not claim the benefit of the 1989 settlement since the company had refused to appear to face criminal charges in India. They concluded instead, ignoring the carefully framed arguments under international law, that only the Indian government had standing to complain of a breach of that settlement because it was the Indian state, not the victims on whose behalf it was supposedly acting, which was a party to that agreement.

On remand from the appellate courts, Union Carbide was obliged, for the first time, to submit to certain limited discovery for documents relating to the history of its operations at Bhopal. One of the documents was a UCC Capital Budget from 1973 for the transfer of technology which Union

Carbide approved to the Bhopal plant for the manufacture of the pesticides including the technology for methyl isocyanate ("MIC"). Under a section entitled "Technology Risks," the document revealed for the first time that the "comparative risk of poor performance and of consequent need for further investment to correct it is considerably higher in the UCIL operation than it would be had *proven technology* been followed throughout," noting in particular that "even the MIC-to-Sevin process, as developed by UCC, has had only a limited trial run." In March of 2004, the same appellate court ruled that the claims of those affected by environmental contamination may be allowed to proceed, including claims for environmental remediation of the land of the Bhopal factory and groundwater aquifer.

One of the imperatives of this anniversary should be to remedy the failure of law which made the Bhopal travesty possible and to prevent its recurrence when the law is again confronted, as it almost certainly will be in this age of globalization, with another disaster having the same international or cross-border dimensions, and the same vexing complexities of liability, corporate structure, jurisdiction, conflicts-of-law and forum. Legal reform is a cause that has seldom, if ever, managed to fire the activist imagination in India or elsewhere. If anything, however, the Bhopal travesty offers a dizzying, vertiginous glimpse of how the economic logic of globalization can become inextricably intertwined with a politics of catastrophe, in which developing countries bargain with foreign investors and multinational corporations by staking the health and lives of their citizens or their environment like gambling chips. This kind of moral abyss is made possible and underwritten by the failures and lacunae in the "rule of law," domestic and international.

The forces that created Bhopal are on the march everywhere today. A recently released report by the United Nations Research Institute for Social Development, called for the implementation of such a legally binding international code of conduct. Some 60,000 multinational corporations in 1998 accounted for more than one-third of world exports, with annual turnovers that dwarfed the gross domestic products of most of their host states in the developing world. "There is a danger that corporate self-regulation, as well as various partnership arrangements," the report warned, giving the example of Bhopal, "are weakening the role of national governments, trade unions and stronger forms of civil society."

As an Organization for Economic Cooperation and Development (OECD) study pointed out, as early as 1993: "Environmentally-dirty industries, particularly resource-based sectors, have migrated over the last two decades to lower income countries with weaker environmental standards; the result is a geographical shift in production capacity within sectors with a consequent acceleration of industrial pollution intensity in developing countries."[4] The study adds that "liberalized trade and investment rules among countries with unequal levels of environmental protection may create incentives for companies to relocate to jurisdictions with lower levels of environmental regulation and lower compliance costs."

The dilemma of law, as Bourdieu has pointed out, is that any rule-based system no matter how impartially administered or fairly conceived will tend inexorably towards outcomes that are not necessarily consonant with intuitive constructions of justice. By the same token, one can scarcely conceive of a political dispensation or social configuration that can be trusted, in and of itself, to deliver any "social justice" worth the name without something closely resembling the normative episteme of the rule of law. Yet, the law cannot remain indifferent to the demands of justice on the specious grounds of an appeal to higher-order concepts of justice "under the law" as the sole and exclusive foundation of its claims to "universal acceptance through an inevitability which is simultaneously logical and ethical," as Bourdieu has written. In the last analysis, the law must secure acceptance of its moral authority from those who seek its protections by aspiring to deliver some modicum of meaningful justice or else depend entirely on the armed power of the state to justify its pretensions to legitimacy.

In the context of Bhopal, that objective requires India and the international community to undertake the following measures to ensure that the law will finally remedy this perversion of justice on its twentieth anniversary.

India must commit itself to legal reform. The Bhopal case presents the spectacle of an official indictment of its own legal system by the country's government before the courts of a foreign power. This is nothing short of an acknowledgment that the sovereign, with full knowledge of its consequences, has deliberately been unwilling or unable to remedy this problem in the more than half-century since independence. Civil litigation in India remains subject to delays of 20 years or more. These kinds of delays are effectively tantamount to a denial of justice. India is a signatory to the International Covenant on Civil & Political Rights which provides, in Article 14, that: "In the determination... of his rights and obligations in a suit at law, everyone shall be entitled to a fair and public hearing by a competent, independent and impartial tribunal established by law."

Legal reforms in India must include provisions for representative suits or class actions to address mass claims of liability like those in Bhopal. Never again should victims be subjected to something like the Bhopal Act which not only enabled the government to function as their lawyer, without observing the minimal professional duties or ethical obligations of an attorney towards his client, but permitted the government to begin functioning as the client as well, stripping the victims of any legal personality and denying them any meaningful role in the decisions that affected their case. Article 16 of the International Covenant guarantees that: "Everyone shall have the right to recognition everywhere as a person before the law."

In honor of the victims of Bhopal, India should lead the effort to enact into international law by treaty or other instrument a legally binding and enforceable code of conduct for multinational corporations, including provisions of liability concerning the export of hazardous technology. Our legal strategy seems to have presaged the UN Human Rights Norms for Business and its Commentary which were recently approved by the Sub-Commission on the Protection and Promotion of Human Rights. These do not, however, have the force of law. India should work to transform these norms into an international legal framework.

The International Law Commission (ILC), an authoritative body charged by the United Nations with the progressive development and codification of international law, has already articulated the majority view that international liability, arising from transboundary environmental harm like Bhopal, should be imposed on states that export hazardous technologies. Even before the disaster, some members of the Commission and the Sixth Committee had suggested that the state of nationality of a multinational corporation should be liable when it "exports" dangerous industries to developing states and harm results. During the discussions on these issues, the U.S. delegation expressed their official view that "under customary international law, States are generally liable for significant transboundary harm caused by private entities acting on their territory or subject to their jurisdiction or control" adding only that "from a policy point of view, a good argument exists that the best way to minimize such harm is to place liability on the person or entity that causes such harm, rather than on the State." The Union of India agreed that liability for environmental harm originating in another state "must be imputed to the operator who was in direct physical control of the activity." But the split in the Commission on this question, ensured that the Commission's draft articles of "International Liability for Injurious Consequences Arising from Acts Not Prohibited By International Law", remain unsettled on this point.[5] India should work to close this omission in the draft articles and resolve the issue of international liability for the export of hazardous or dangerous technology.

Last but not least, India must secure the appearance of UCC or its new parent, Dow Chemical, to face trial on the criminal charges pending against it in the Bhopal District Court. The criminal case against Indian officials has dragged on for a number of years and judges presiding over the case against UCC have been repeatedly transferred. A single judge should be appointed to preside over the entire matter and expedite proceedings so that the criminal case can be quickly adjudicated and disposed of under the law. India has an obligation to ensure that this crime is effectively prosecuted. Article 8 of the Universal Declaration of Human Rights stipulates that: "Everyone has the right to an effective remedy by the competent national tribunals for acts violating the fundamental rights granted him by the constitution or by law." Pursuant to Article 2 of the International Covenant on Civil & Political Rights, India has undertaken to "ensure that any person whose rights or freedoms as herein recognized are violated shall have an effective remedy," and to "ensure that the competent authorities shall enforce such remedies when granted."

Footnotes

1 Pierre Bourdieu, "The Force of Law: Towards A Sociology of the Juridical Field," Hastings Law Journal 38 (5): 814-53 (1987).

2 *Union Carbide Corp. v. Union of India* , 1 S.C.C. 674, 675 (1989).

3 *Union Carbide Corp. v. Union of India* , Review Petition Nos. 229 and 623-24 of 1989, 70 (1991).

4 Candice Stevens, Synthesis Report: Environmental Policies and Industrial Competitiveness, in Environmental Policies and Industrial Competitiveness 7 (OECD ed., 1993).

5 International Law Commission, Second Report on International Liability for Injurious Consequences Arising out of Acts Not Prohibited by International Law (Prevention of Transboundary Damage from Hazardous Activities, A/CN.4/501, 5 May 1999.

Source: from *Seminar 544: Elusive Justice*, December 2004, New Delhi, India.

Troubled compensation (2004)

On the twentieth anniversary of the disaster, Amnesty International issued a report reviewing the responses to the Bhopal disaster. In this section describing the compensation process, the report articulates some of the hurdles that have come between claimants and settlement money.

The May 1989 order of the Supreme Court setting out the settlement stated: "No individual claimant shall be entitled to claim particular quantum of compensation even if his case is found to fall within any of the broad categories."[250] This meant that victims were denied their individual right to prove the extent of individual damages suffered and claim appropriate compensation.

Claims were adjudicated in claims courts by Claims Commissioners, Additional Claims Commissioners and the Welfare Commissioner (a sitting judge of the Madhya Pradesh High Court).[251]

Claimants had to pass through several stages in order to secure compensation: registration; identification (requiring proofs of identity, residence and medical records to prove gas effects); notification of their hearing; categorization; adjudication and, for an unfortunate few, the appeals process.

Survivors say that the process involved innumerable trips to hospitals, government offices, lawyers, banks and the court. They said they had to stand for hours in long lines and endure apathy, indifference, suspicion and corruption at the hands of employees, brokers, middlemen and lawyers. For poor and illiterate people, the process was fraught and frustrating, and at the end they gained very little

A 1995 assessment reveals that the maximum average compensation was awarded not in the two severely affected wards (Nos. 13 and 20) but in Ward No. 21, categorized as mildly affected.[252] In 1995 the average compensation received for personal injury was Rs.26,531, just above the stipulated minimum of Rs.25,000 (around $545 at current rates). Average awards were far smaller than originally envisioned. This indicates the arbitrary nature of the awards. The difference between the highest and the lowest average compensation paid for injury was Rs.8,483, although the 1992 guidelines issued to the Welfare Commissioner stated that the compensation for injuries should be in the range Rs.25,000 to Rs.400,000 (around $8,700). In at least five wards the average compensation was equal to the minimum, while in six wards it was actually less than the minimum. In cases where the victim had died, the average compensation given by 1995 was Rs.73,638 (around $1,605), far less than the minimum Rs.100,000 stipulated by the Supreme Court. An October 2002 survey in one severely affected ward revealed that 91% of the 1,481 claimants had received only the minimum compensation.[253]

Fast-track justice?

In 1995 special fast-track courts called Lok Adalats were set up to expedite the processing of thousands of claims in the claims courts. The lack of due process in these courts was described by a committee appointed by the Supreme Court:

> "In the Lok Adalats, a particular amount was specified and the claimants were coerced to accept this amount and accord their consent to the medi-

cal categorization... In the office of the Lok Adalats, no legal assistance was available to the claimants."[254]

Lawyers and counsel were barred from representing victims in Lok Adalats. Victims were completely unaware of the process, and directions on minimum compensation were not followed. Claimants had to go to the Supreme Court to secure the right to appeal against the decisions of the Lok Adalats.[255]

Delays

Claimants faced significant delays at every stage of the process, even after adjudication. "The cheque was awarded at least two months after the judgment. And the money was available only a further 40 days after the award," said Shanti Devi, herself a victim and now an activist.

Delays were aggravated by the fact that claimants were not paid any interest for the delay on the amounts payable to them. Interim relief of Rs.200 per month was ordered by the Supreme Court in March 1990 because adjudication of claims had not started. This was deducted from the eventual compensation that victims secured.

Problems with medical categorization

The Process of Injury Evaluation (PIE) categorized the degree of disability or injury according to scores given to symptoms, signs, treatment received and investigation results. Evidence suggests that claims of medical injury were not accurately scored.

The PIE relied mostly on three investigations: X-rays, the Pulmonary Function Test (PFT) and the Exercise Tolerance Test (ETT). However, these were not widely administered: a 1989 study showed that while at least 60% of the victims required PFT and ETT, the claims directorate had ordered only 15% and 2% respectively to take these tests. The state government declared that "it was not practicable to subject every claimant to these time-consuming investigations in mass operations like this."[256]

The medical records and the PIE did not assess how victims' exposure and subsequent illness affected their ability to carry out their normal level of activities and their work. The ability of a claimant to produce medical records for the post-exposure period was critical. "A large number of victims were being categorized as 'no injury' even though they are ill and can produce proof of residence in the exposed area, all because they cannot produce medical documents for the post-exposure period."[257]

As a result of the paucity of quality medical research on the consequences of the Bhopal gas leak and lack of comprehensive information about the toxicity of MIC, neither those claiming to have been affected nor those adjudicating their claims have had any rigorous basis to understand the link between the exposure to gas and the origin of health-related disabilities. This has given rise to a widespread sense of grievance that compensation has been arbitrarily decided.

Failure to register claims

A study by the Bhopal Group for Information and Action of three gas-affected localities concluded that the claims of 42.4% of the residents had not been registered. In one severely affected locality, nearly one sixth of the claims were not registered. The single largest omission comprised at least 15,000 gas-affected victims who were under 18 at the time of registration of claims. Not until August 1992 did the Supreme Court order that minors had a legal entitlement to be registered. Children born to gas-affected parents have continued to be excluded, despite the Supreme Court recognizing the entitlement of "later born children who might manifest congenital or pre-natal MIC afflictions."[258]

Failure to pay the compensation available

Of the Rs.750 crores (around $470 million at the prevailing rate) settlement, UCC contributed $420 million, which was held in a US dollar account, and UCIL contributed Rs.68.99 crores (around $44 million), held in a rupee account.[259] The money was available in 1989 but the claims courts began adjudicating cases only in 1992 and the process is still not complete.

Over the years, owing to the appreciation of the US dollar vis-a-vis the Indian rupee and the interest earned on undistributed funds, the sums held have grown considerably.[260] As of mid-2004, a total of Rs.1,503 crores ($327.5 million) was held by the Reserve Bank of India and Rs.1,535.58 crores ($334.6 million) had been disbursed by claims courts.[261]

After approaches by victims' groups, on 19 July 2004 the Supreme Court ordered the disbursal of the remaining funds, giving each of approximately 570,000 victims the same amount they had earlier received as compensation.[262]

Corruption

The claims system saw hundreds of thousands of poor and illiterate survivors facing a complex bureaucratic system. Survivors complain that the system required excessive paper work and complicated procedures and that this opened the way for intermediaries, brokers and opportunistic lawyers. Nanni Bai, a widow, paid Rs.60,000 to a lawyer and broker to procure compensation of Rs.100,000 for her husband's death. Ahmadi Bai, 65, paid Rs.500 to a doctor to testify that her illness was because of her exposure. A number of survivors say that even the person who delivered the notification of the date of the claim hearing had to be bribed.

Kiran Jain, a 40-year-old widow, said: "Having all your papers is not enough. You have to pay a bribe for everything even to get a Pension Book or a Below Poverty Line card. If you pay, you get what you want; if you don't, then you just suffer."[263]

Footnotes:

250. Order 05-04-1989 in Civil Appeal Nos. 3187-89, *Union Carbide Corporation v Union of India*, Supreme Court of India.
251. Ramanathan, Usha, "A Critical Analysis of Laws Relating to Personal Injury," dissertation submitted to Delhi University, September 2001.
252. "Dismal State of Disbursal of Compensation to Victims of Union Carbide Gas Disaster," Bhopal Group for Information and Action, 1996.
253. "Survey of compensation among residents of Jai Prakash Nagar," Documentation Unit, Bhopal Peoples' Health & Documentation Clinic, Sambhavna Trust, Bhopal, 2002.
254. "A Critical Analysis of Laws Relating to Personal Injury," *op cit.*
255. "A Critical Analysis of Laws Relating to Personal Injury," *op cit.*
256. "Critique Of Medical Categorization: The Process of Injury Assessment Followed by the M.P. Government is Faulty," Dr Nishith Vohra and Dr Sathyamala, 26 December 1989.
257. "Critique Of Medical Categorization," *op cit.*
258. "Compensation Disbursement, Problems and Possibilities, A Report of a Survey Conducted in Three Gas Affected Bastis of Bhopal," Bhopal Group For Information And Action, January 1992.
259. Order 15-02-1989 in Civil Appeal Nos. 3187-89, *Union Carbide Corporation v Union of India*, Supreme Court of India.
260. The value of the US dollar has risen from an average of Rs.14.48 in 1988-89 to Rs.48.39 in 2002-03, an increase of some 350%.
261. Order 19-07-2004 in IA Nos 46-47 Civil Appeal Nos 3187-89, *Union Carbide Corporation v Union of India*, Supreme Court of India.
262. Order 19-07-2004, *op cit.*
263. A number of other victims and survivors as well as organizations and activists working with survivors confirmed this.

Source: *Clouds of Injustice: Bhopal, 20 years on,*
Amnesty International report, 2004.

5

Criminals – and Corporations

Why has no one been held accountable?

THE OCTOBER 1991 Supreme Court decision upholding $470 million as "full and final" for all damage and human injury claims by the survivors against Union Carbide also restored the criminal charges of culpable homicide against Warren Anderson, Chairman of Union Carbide Corporation at the time of the disaster, several officials of Carbide's Indian subsidiary, Union Carbide Eastern, Hong Kong, Union Carbide India Ltd. (UCIL), and the parent U.S. corporation. Many critics of the Supreme Court felt the charges should never have been quashed in the first place. But a central issue in pursuing the criminal charges against Warren Anderson and the Union Carbide Corporation remains whether or not they are subject to the criminal law jurisdiction of the Indian courts. Carbide and Anderson insist that, even if they had been, they no longer are.

Bhopal unfortunately demonstrates the utter inadequacy of the development of corporate criminal law at a time when a corporate crime wave appears to be sweeping across the globe. Despite the increasingly transnational nature of corporate crimes, corporations remain above the law, and their officers are not held accountable for their actions. Warren Anderson, for example jumped bail in India in 1984. In 2002, a muckraking paper in the U.K. found him living in the Hamptons, near New York, USA still receiving his retirement checks. While India has recently delivered his arrest warrant to the United States Justice Department, it has declined to extradite him on specious grounds.

It is as a result of these processes that shunt responsibility between parties and countries, and water down the charges at every stage, that the criminal prosecution of individuals and corporations in the Bhopal case has not advanced much from 1984. The history of the criminal case is replete with evidence of lack of political will of the Indian government, arbitrary decisions by the highest court of the land and the incompetence of the prosecuting agency – the Central Bureau of Investigation [CBI]. These include the dilution of criminal charges against the accused Indian officials from culpable homicide to negligence, de-registration of United Carbide Eastern, Hong Kong, a wholly owned subsidiary of UCC which subsequently disappeared into thin air, and the sale of shares of UCC that were confiscated as a means to ensure the presence of the absconding corporation in the criminal court. The silver lining in the bleak history of criminal justice in Bhopal is provided by the Chief Judicial Magistrates – the men lower down the judicial totem pole. It is primarily their orders and directions that have helped keep the legal case alive, the most recent being that of 6 January 2005 requiring the Dow Chemical Company, USA to state why it should not be asked to present absconding accused Union Carbide Corporation in the ongoing criminal case on the disaster in Bhopal.

But Warren Anderson himself has become the figurehead for impunity in Bhopal. Every year he is burned in effigy during street protests. The phrase "Hang Warren Anderson" appears in slogans and graffiti in Bhopal and beyond. He was the man who said that the corporation accepted "moral responsibility" for the disaster, a meaningless fig leaf as this perplexing position does not admit legal liability, and a corporation has no properly "moral" conscience (despite its legal status as a "person"). Bhopal is a classic tragedy, and Anderson, even if he had no ill intent, is the responsible party and must be punished for not rectifying or preventing the unstable situation in Bhopal about which he *knew, or should have known.* From the original charge sheet, through the shell game of corporate divestment and name changes, this chapter tracks the effort to bring the parties responsible for the Bhopal disaster to trial.

CBI charge sheet for the Bhopal gas disaster (1984, 1987)

In the immediate aftermath of the disaster, India's Central Bureau of Investigation (CBI) enumerated the criminal acts alleged to have caused the Bhopal disaster. Twenty years later, all of these crimes remain unpunished.

Station SPF-ACU-I New Delhi-District Delhi
Charge sheet No. 2
Dated 30 November 1987
Address and Occupation of Complainant or informant:
Shri S. S. Thakur, T. I. Hanumanganj, Bhopal. First information report number 3 Date 6th December 1984.

1. Shri Warren Anderson, Former Chairman,
   The Union Carbide Corporation Old Ridgebury Road,
   Danbury, Conn. USA.
2. Shri Keshub Mahindra, Former Chairman,
   UCIL 15, Mathew Road, Bombay – 400 004 Residing at Flat No. 9 & 10,
   St. Helen's Court, G. Deshmukh Marg, Bombay – 400 026.
3. Shri Vijay Gokhale, former, Managing Director and presently
   Chairman-cum Managing Director,
   15, Mathew Road, Bombay – 400 004.
4. Shri Kishore Kamdar, former Vice-President In charge,
   A.P. Division UCL 15, Mathew Road, Bombay – 400 004 residing at
   Kshitij, 19th Floor, Flat No.191, Napean Sea Road, Bombay – 400 006
5. Shri J. Mukund former works manger, A.P. Division Bhopal,
   residing at 6 D Lands End, Downgersi Road, Bombay – 400 006
6. Dr. R. B. Roy Chawdhury, former A Asst. Works Manager,
   A.P. Division UCIL. Berasia Road, Bhopal, residing at Satya
   Flat No. 10, 15th Road, Bandra, (W) Bombay – 400 050.
7. Shri S. P. Choudhury, former Production manager
   A.P., Division , UCIL Berasia Road, Gultakdi, Pune – 410 037.
8. Shri K. V. Shetty, Plant Superintendent,
   A.P. Division, UCIL, Berasia Road, Bhopal.
9. Shri S. I. Qureshi, Production Assistant,
   A. P. Division UCIL, Berasia Road, Bhopal
10. Union Carbide Corporation
    39, Old Ridgebury Road, Danbury, Connecticut, USA, 06817
11. Union Carbide (Eastern) Inc.
    16th Floor, New World Office Building (East Wing)
    24 Salisbury Road, Tismsha Tsui, Knowles Hong Kong.
12. Union Carbide India Ltd., I, Middleton Street, Calcutta – 16.

Further list, if any will be submitted on completion of remaining investigation.

Dispatched at 6.45 p.m.
Signature of Investigation Officer
On 30th November, 1987

1. Union Carbide India Ltd., the majority shareholding of which is held by UCC, USA, was running a factory at Bhopal for the manufacture of pesticides. The main chemical from which the pesticide Sevin was manufactured was Methyl Isocyanate ($CH_3N=C=O$) which was also being manufactured in the same factory and was being stored in underground tanks. The factory is presently not functioning.

2. On the night of 2nd December, 1984 from about 00.00 to 00.45 hrs (on 3rd December 1984) onwards, MIC started to escape from tank No. 610 in the factory in large quantities causing the death of thousands of human beings (Sic illegible) animals and injuring also the health of many thousands of human beings and animals on short as well as long-term basis.

3. Crime No. 110/84 registered at police station Hanumanganj at Bhopal on 6th December 1864 by S. H. Shri Surinder Singh Thakur, the Inspector who observed people dying around the factory of Union Carbide India Ltd., Bhopal (UCIL) due to escape of some gas from the factory. He registered the case *suo moto* under section 304A IPC. There was no information available at that time from anyone in the Factory. But on enquiries made by him during the course of the day, five employees of the Factory (A5 to A9) were arrested and kept in police custody. Accused No.1 Shri Warren Anderson was arrested along with accused No. 2 and 3 on 7th December 1984. Shri Warren Anderson was released on bail the same day by the I. O. after completing the required legal formalities, C.B.I. (D.S.P.E.) registered a case on 6th December 1984 as RC – 3/84-CIU (I) U/s. 304 A IPC and received the records of the case from the local police on 9th December 1984 along with A2, A3, and A5 to A9 in police custody from the Madhya Pradesh Police.

4. Investigation has revealed that the Union Carbide Corporation is a company with headquarters in U.S.A. having affiliate and subsidiary companies throughout the world. These subsidiaries were supervised by four regional offices which were controlled by UCC, U.S.A. UCIL is a subsidiary of UCC, U.S.A. Union Carbide Eastern Inc. with its office in Hong Kong is the regional office of UCC, U.S.A. which controlled UCIL, India, besides others. UCC, U.S.A. got incorporated in India on 20th June 1934 a Company known as the Eveready Company (India) Ltd., under the Indian Companies Act (Act VII) of 1913 with the Registrar of Joint Stock Companies, Bengal. The name of the company was further changed w.e.f. 24th December 1959 into Union Carbide India Ltd. Under the Indian

Companies Act, 1956. The UCC was a majority shareholder (50.9 %) in UCIL. UCC nominating its own Directors to the Board of Directors of the UCIL and was exercising strict financial administrative and technical control on the Union Carbide India Ltd. Thus all major decisions were taken under the orders of the Union Carbide Corporation of America. The evidence collected during the investigation proves that UCC was in total control of all the activities of UCIL.

5. The investigation of this case was dependent on highly scientific and technical evaluation of the events which led to the escape of MIC gas from the UCIL plant at Bhopal. The government of India therefore constituted, immediately after the incident a team headed by Dr. S. Varadarajan, then DG/CSIR to study all the scientific and technical aspects and submit their report. Dr. M. Sriram, Chief Research and Development Manager, Hindustan Organic Chemicals, Rasayani, District Raigad (Maharashtra), was a member as well as coordinator of the scientific team. Dr. Varadarajan submitted the report in December, 1985. A further back up report was submitted by the CSIR in May 1987. These reports furnish, *inter alia*, the causes that fed to the incident.

6. Investigation has revealed that UCIL started importing Sevin from the UCC, USA in December 1960, They were marketing this Sevin after adding dilutants etc. Subsequently they decided to manufacture Sevin in their plant at Bhopal itself and accordingly created necessary facilities for production of Sevin with MIC as the basic raw material. To start with, they were importing MIC in 200 liters capacity stainless steel drums from the UCC plant in West Virginia USA. Subsequently, UCC and UCIL decided to manufacture MIC in their factory at Bhopal. Itself.

7. At that stage on 13th November, 1973, UCC and UCIL entered into an agreement entitled Foreign Collaboration Agreement according to which the best manufacturing information then available from or to Union Carbide to be provided for the factory in India. This necessitated UCC supplying the design, know-how and safety measures for the production, storage and use of MIC which ought to have been an improvement on the factory of UCC at West Virginia based on the experience gained there. Investigation has however disclosed that the factory at Bhopal was deficient in many safety aspects. The design, know-how and safety measures were provided by the Union Carbide Corporation, USA and the erection and commissioning of the plant was done under the control of the experts of UCC. The Indians in this plant were only working under their directions.

8. After an initial period of profit, the UCIL factory was running in loss. The loss for the first 10 months of 1984 amounted to Rs.5,03,000. Due

to this, U.C.E. Hong Kong directed UCIL vide their letter dated 26th October, 1984 that the factory at Bhopal should be closed down and sold to any available buyer. As no buyer became available in India, UCE Hong Kong, directed UCIL to prepare an estimate for dismantling the factory and shipping it to Indonesia or Brazil where they probably had some buyers. These estimates were completed towards the end of November 1984.

9. The investigation conducted by the C.B.I., the report of the scientific team established by Government of India and in particular the literature and manuals etc. regarding MIC of Union Carbide Corporation itself prove that MIC is reactive, toxic, volatile and flammable. It is a highly hazardous and lethal material by all means of contact and is a poison. Skin contact with MIC can cause sever burns. MIC can also seriously injure the eyes even in 1% concentration. Exposure to MIC is extremely, irritating and would cause chest pain coughing, choking and even pulmonary edema. On thermal decomposition, MIC would produce hydrogen cyanide, nitrogen oxide, carbon monoxide and/or carbon dioxide.

10. MIC has to be stored and handled in stainless steel of types 304 or 316 namely good quality stainless steel. Using any other material could be dangerous. In particular, iron or steel, aluminum, zinc or galvanized copper, or tin or their alloys could not be used for purposes of storage, transfer/ transmission of MIC. This would mean that even the pipes and valves carrying MIC had also got to be of the prescribed stainless steel. In other words, at no stage should MIC be allowed to come into contact with any of the metals mentioned above.

11. The tanks storing MIC have to be for reasons of safety, twice the volume of the MIC to be stored. It was also advised by UCC itself that an empty tank should also be kept available at all times for transferring MIC from its storage tank to standby tank on occasions of emergency. MIC has to be stored in the tanks under pressure by using nitrogen which does not react with MIC. The temperature of the tanks with MIC has to be maintained below 15 °C and preferably at about 0 °C. The storage system and the transfer lines have to be free of any contaminants as even trace quantities of contaminants are sufficient to initiate reaction which could become runaway reaction on reaction setting in, there could be dangerous and rapid trimerization. The Induction period could vary from several hours to several day. The heat generated could cause reaction of explosive violence. In particular, water reacts exothermically to produce heat and carbon dioxide. Consequently, the pressure in the tank will rise rapidly if MIC is contaminated with water. The reaction may begin slowly, especially if there is no agitation but it will become violent. UCC itself states that with bulk

systems contamination is more likely than with tightly sealed drums. All these properties of MIC show that despite all the safety precautions that could be taken, storage of large quantities of MIC in big tanks was fraught with considerable risk.

12. Investigation has disclosed that at the time the incident took place there were three partially buried tanks in the factory at Bhopal. These were numbered E610, E 611 and E 619. MIC was being stored generally in the tanks E610 and E611. E619 was supposed to be the stand by tank. In the normal running of the factory, MIC from E610, and 611 was being transferred to the Sevin plant through stainless steel pipe lines. MIC is kept under pressure by nitrogen which is supplied by a carbon steel header common to all the storage tanks. There is a strainer in the nitrogen line. Subsequent to the strainer the pipe is of carbon steel and leads to make up control valve (DMV) which also has a body of carbon steel. These carbon steel parts could get exposed to MIC vapors and get corroded, providing a source of contaminant which could enter the MIC storage tank and cause dangerous reactions in the MIC. During the normal working of the factory, MIC fumes and other gases that escape pass first through a pipe line called Process Vent Header (PVH) of 2" diameter. The escaping gases were carried by the PVH line to a Vent Gas Scrubber (VGS) containing alkali solution which would neutralize the escaping gases and release them into the atmosphere. Another escape line of such gases that was provided from the Relief Valve Vent Header (RVVH) of 4" diameter. Normal pressure of the MIC tank is shown by a pressure indicator. When the pressure in the tanks exceeded 40 psig, a rupture disc (RD) leading to a safety relief valve (SRV) had to break and the said SRV in the RVVH line open automatically to allow the escaping gas to travel through the RVVH line to the VGS for neutralization.

13. Investigation has shown that the PVH and RVVH pipe line as well as the valves therein were of carbon steel. Besides, on account of design defect these lines also allowed back flow of the alkali solution from the VGS to travel up to the MIC tanks.

14. Very essential requirement was that the MIC tanks in the factory had to be kept under pressure of the order of 1 kg/cm^{2}g by using nitrogen, a gas that does not react with MIC. However, MIC in tank No. 610 was stored under nearly atmospheric pressure from 22nd October, 1984 and attempts to pressurize it on 30th November and 1st December 1984 failed. The design of the plant ought not have allowed such a contingency to happen at all. The tank being under nearly atmospheric pressure, free passage was available for the entry of back flow of the solution from the VGS into the tank. According to the report of Dr. Varadarajan Committee, about 500 kgs water with contaminants could enter tank 610 through RVVH/PVH

lines. The water that entered RVVH at the time of water flushing along with backed up alkali solution from the VGS already present could find its way into the tank 610 through the RVVH/PVH lines via the blow down DMV or through the SRV and RD.

15. The first indication of any reaction in the tanks comes through the pressure and temperature indicators. The thermowell and temperature transmitting lines were out of order throughout and no temperature was being recorded for quite some time. Pressure was also being recorded at the end of each shift of 8 hours duration instead every 2 hours as was being done earlier.

16. Shifts in factory ended at 6.45 a.m, 2.45 p.m. and 10.45 p.m.

17. On 2nd December, 1984 before 10.45 p.m. one deviation was noticed in the pressure of tank no. 610. Soon thereafter, some operators noticed leakage of water and gases from the MIC structure and they informed the Control Room. The Control Room operator saw that the pressure had suddenly gone up in tank No. 610. Some staff in the third shift including S/ Shri, R. K. Kamparia. C. N. Sen and Saumen Dey checked the pressure indicators on the tanks E610 and found that the pressure had gone out of range. The factory staff tried to control the situation but even tank E619 which had to be kept empty for emergency transfer was found to contain MIC and when the reaction started, transfer thereto from tank 610 was not possible. The staff on duty immediately informed senior officials of UCIL at Bhopal about the escape of MIC. During all these developments and even thereafter the Union Carbide officials at Bhopal did not give any information to the residents or any local authority about the serious dangers to which the people were exposed and regarding which these said officials had full knowledge. On the other hand, what was initially mentioned, was that ammonia gas had escaped.

18. The scientific team headed by Dr. Varadarjan has concluded that the factors which led to the toxic gas leakage causing its heavy toll existed in the unique properties of very high reactivity volatility and inhalation toxicity of MIC. The needless storage of large quantities of the material in very large size containers for inordinately long periods as well as insufficient caution in design, in choice of materials of construction and in provision of measuring and alarm instruments, together with the inadequate controls on systems of storage and on quality of stored materials as well as lack of necessary facilities for quick effective disposal of material exhibiting instability, led to the accident. These factors contributed to guidelines and practices in operations and maintenance. Thus the combination of conditions for the accident were inherent and extant.

19. Post mortem, medical and other evidence prove that the deaths and injuries were caused due to the exposure of the people to MIC and its derivatives, including cyanide.

20. The investigation conducted by the C.B.I. has proved the following:

I MIC is a highly dangerous and toxic poison.

II. Storing huge quantity of MIC in large tanks was undesirable and dangerous as the capacity and actual production in the Sevin plant did not require such a huge quantity to be stored. Only adequate quantity of MIC should have been stored, that too in small separate stainless steel drums.

III. The VGS that had been provided in the design was capable of neutralizing only 13 tones of MIC per hour and proved to be totally inadequate to neutralize the large quantities of MIC that escaped from tank No. E 610. When the two tanks (610 and 611) themselves had been designed for storing a total of about 90 tons of MIC proportionately large capacity VGS should have been furnished in the design and erected rather than the VGS that was actually provided.

IV. Due to the design defect, there was back flow of alkali solution from the VGS to the tanks which had been drained in the past by the staff of UCIL. In fact, even after the incident such draining was done from the PVH and RVVH lines.

V. Whereas the MIC tanks had to be constantly kept under pressure using nitrogen the design permitted the MIC tanks not being under pressure in certain contingencies.

VI. The refrigeration system that had been provided was inadequate and inefficient. No alternate standby system was provided.

VII. Neither the UCC nor the UCIL took any steps to apprise the local administration authorities or the local public about consequences of exposure of MIC or the gases produced by its reaction and the medical steps to be taken immediately.

21. Apart from these design defects the further lapses that were committed were:

a) Invariably storing MIC in the tanks which was much more than the 50% capacity of the tanks which had been prescribed.
b) Not taking any adequate remedial action to prevent back flow of solution from VGS into the RVVH and PVH lines. This alkali solution/water, therefore, used to be drained.

c) Not maintaining the temperature of the MIC tanks at the preferred temperature of 0 °C but at ambient temperatures which were much higher.
d) Putting a slip blind in the PVH line and connecting the PVH line with a jumper line to the RVVH line.
e) Not taking any immediate remedial action when tank No. E 610 did not maintain pressure from 22nd October, 1984 onwards.
f) When the gas escaped in such large quantities, not setting out an immediate alarm to warn the public and publicize the medical treatment that had to be given immediately.

22. Investigation has shown that even if these lapses had not occurred, still the incident would have taken place due to the basic defects in the design supplied by UCC whose experts supervised the erection and commissioning of the plant itself . The lapses only helped to aggravate the consequences of the incident. The lapses were also of such nature which could be reasonably foreseen as inevitable in any such factory and the design ought to have acted to ensure total safety even if such lapses took place. The design did not however do so.

23. The evidence collected during the investigation proves that the accused persons had the knowledge that by the various acts of commission and omission in the design and running of the MIC based plant, death and injury of various degrees could be caused to a large number of human beings and animals. All the accused persons joined in such acts of omission and commission with such common knowledge. This resulted in the incident on the night of 2nd / 3rd December, 1984 which caused the death immediately and till date of about 2850 human beings and about 3000 animals. The number of affected persons is more than 5,00,000. The ailments developed by the affected persons include damaged respiratory tract function, gastro intestinal functions, muscular weakness, forgetfulness etc.

24. The investigation has established that S/Shri Warren Anderson, then Chairman, Union Carbide Corporation, USA; Keshub Mahindra, then Chairman, UCIL Bombay: Vijay Gokhale, then Managing Director and presently Chairman cum Managing Director UCIL, Bombay; Kishore Kamdar, then Vice-President in charge, A.P. Division UCIL Bombay; J. Mukund, then Works Manager, A.P. Division UCIL, Bhopal K.V. Shetty, Plant Superintendent, A.P. Division, Bhopal; S. I. Qureshi, Production Assistant, A.P. Division, UCIL Bhopal: the Union Carbide Corporation, U.S.A.; Union Carbide Eastern Inc. Hong Kong and Union Carbide India Limited Calcutta have committed offences punishable under section 304. 326, 324, 429, IPC r/w section 35 IPC.

25. Due to the complicated nature of the case and certain difficulties that were encountered in the investigation some further investigation still remains to be done which is proposed to be continued after submission of this charge sheet. While the control exercised by Union Carbide Eastern Inc. Hong Kong, over UCIL has been proved during the investigation by the records of UCIL, this requires to be further confirmed by interrogating the concerned executives of the Hong Kong company and collecting the relevant documents. The same has to be done in respect of UCC, USA also. In particular, the UCC plant at West Virginia in USA has also to be inspected. Some further investigation is also to be done with reference to the records of the Government of Madhya Pradesh many of which are still to be made available to the Investigation Officer. Though such further investigation is statutorily permitted under section 173 (8) of the Code of Criminal Procedure, 1973, it is requested that this Hon'ble Court may be pleased to take note of this fact and permit the same.

26. It is therefore, prayed that this Hon'ble Court may summon the accused persons and conduct the trial according to law in respect of the offences mentioned above.

(B. K. Shukla)
Dy. Supdt. Of Police
CBI : Acu (I): New Delhi

Encl: -

1. "Jamanat Nama" of Shri W. M. Anderson, dated 7th December, 1984. Surety given by Shri A.M. Kuruvila, then General Accountant. UCIL, Bhopal.
2. "Muchalaka" of Shri Warren Anderson, dated 7 December, 1984.
3. List of documents (28 sheets)
4. List of witnesses. (10 sheets)

Submitted to court by
U.S. Prasad
Senior Public Prosecutor
C.B.I. A.C.U.I. New Delhi.

Dated 30.11.1987
[The enclosures have been excluded – Eds.]

Source: Sambhavna Clinic Documentation Center, Bhopal.

Warren Anderson's bail bond and arrest warrant (1984–1992)

In 1984 Warren Anderson, then CEO of Union Carbide Corporation, was arrested in Bhopal for culpable homicide in the Bhopal case. He signed a bail bond, paid Rs.25,000, and promised to return for trial. He has still not responded to the warrant issued (copy below).

Warren Martin Anderson

Born 29 November 1921. 6' 2" 210 pounds. Color of hair – White

Majored in Chemistry in 1942 from Colgate. Joined Navy, discharged in 1945. Studied law.

Wife Lillian Anderson, former school teacher, no children.

Joined Union Carbide Corporation in 1945 as salesman and became its President in 1979 and Chairman in 1982. Retired in 1986.

Source: Warren Anderson: A public crisis, a personal ordeal.
The New York Times, May 19, 1985

COPY OF ANDERSON'S BAIL BOND [Translated from Hindi]

Bond

December 7, 1984

I, Warren M Anderson s/o John Martin Anderson am resident of 63/54 Greenidge Hills Drive, Greenidge, Connecticut, USA. I am the Chairman of Union Carbide Corporation, America. I have been arrested by Hanumanganj Police Station, District Bhopal, Madhya Pradesh, India under Criminal Sections 304 A, 304, 120 B, 278, 429, 426 & 92. I am signing this bond for Rs. 25,000/- and thus undertaking to be present whenever and wherever I am directed to be present by the police or the Court.*

Signed : Warren M. Anderson

Note:

Mr. Anderson's signature was obtained after the language of this bond was translated into English by Mr. Gokhale and read out to him.

* Anderson is charged under Indian Penal Code sections 304 [culpable homicide, punishable by 10 years to life imprisonment and fine], 320 [causing grievous hurt punishable by 10 years to life imprisonment and fine], 324 [causing hurt, punishable by 3 years imprisonment and/or fine] and 429 [causing death and poisoning of animals, punishable by 3 years imprisonment and/or fine]

NON-BAILABLE WARRANT

Warrant of Arrest

(Section 70 of the Code of Criminal Procedure, 1973)
(Act II of 1974)
In the court of Chief Judicial Magistrate, Bhopal, Madhya Pradesh, India.
Docket RT No.2792 of 1987.

To,

The Superintendent of Police
Central Bureau of Investigation
Special Police Establishment
Anti-Corruption Unit-I
CGO Complex
Lodhi Road New Delhi

Whereas Mr Warren Anderson, son of Mr John Martin Anderson, former Chairman, Union Carbide Corporation, 39 Old Ridgebury Road, Danbury, Connecticut, USA, 06817, r/o 54 Greenwich Hill Drive, Greenwich CT 06831, USA, forwarding address PO Box 281, Ocean Road, Bridge Hampton, New York 11937 stands charged with the offence under Section 304, 324, 326, 429 IPC r/w Section 35 IPC, you are hereby directed to arrest the said Warren Anderson and to produce him before me. Herein fail not.

Dated: This 10th day of April 1992.

Gulab Sharma
Chief Judicial Magistrate
Bhopal, Madhya Pradesh
India

Wanted – for 20,000 deaths at Bhopal (2002)

The UK's Daily Mirror tracked down Warren Anderson living out a luxurious retirement. The company had claimed they did not know his location.

A FUGITIVE boss wanted for the killing of up to 20,000 people in the Bhopal gas disaster has been tracked down by the Mirror.

EXPOSED: Warren Anderson at his luxury home

Ex-Union Carbide chief Warren Anderson, 80, has evaded the law for 18 years after jumping bail on charges of culpable homicide following the 1984 tragedy in India.

Today, he is living in a £680,000 US holiday home while 120,000 poverty-stricken Bhopal victims are ravaged by health problems and sackfuls of skulls bear testimony to the world's worst industrial disaster.

But time is running out. Yesterday an Indian court ruled Anderson must stand trial for homicide after rejecting a plea that the charges against him be reduced to rash negligence. Lawyers will apply for an arrest warrant to be given to US authorities to carry out. Anderson could be jailed for 10 years.

Confronted by the Mirror at his home in the exclusive Hamptons, outside New York, he was clearly discomfited.

Asked if he felt any responsibility for death and suffering on an awesome scale, he barked "No comment."

Anderson and wife Lillian spend the summer at Bridgetown in the Hamptons and the winter in Miami.

They live in a four-bedroom, four-bathroom house.

Membership of the local tennis club costs £1,750 a year.

It is a far cry from Bhopal where thousands of diseased and deformed victims still pay the price for Anderson's cost cutting.

The boss was chairman of Union Carbide whose pesticide plant in Madhya Pradesh used deadly methyl isocyanate – MIC.

Early on December 3, 1984, five tons of MIC leaked out killing 8,000 on the spot.

Thousands more died later and further thousands are still condemned to a life of unending suffering.

Anderson was later accused of cutting safety to save £30 a day. The crumbling factory still bleeds poison.

Amazingly the Indian government applied to reduce charges against Anderson.

It led to huge protests and claims that the country did not want to scare off investors.

But yesterday the bid was thrown out by a district judge. Bhopal lawyer Raj Punjwani said: "We won't rest until he's here. This is a good day for us."

Anderson retired from Union Carbide, which is now owned by Michigan-based Dow Chemicals, in 1986.

Source: *The Daily Mirror*, UK, August 29, 2002 by Rosa Prince

Is Dow liable for Carbide's misdeeds? (2003)

Bhopal activist Tim Edwards explores Dow's legal responsibility for Bhopal.

In respect of Carbide's liabilities, Dow has been playing a game of Chinese Walls. It's 100 % owner of Carbide but calls itself a 'shareholder'. Carbide's CEO is Dow's head of Catalysts. One of Carbide's two directors is Dow's European CEO. Carbide's SEC registrar is Dow's Frank H Brod. No doubt very elaborate and hyper technical legal distinctions were drawn up between the two companies at the time of the merger, but the market for one isn't buying it: when Carbide's asbestos liabilities came to light it was Dow's share price that took a rocket. According to Andrew Brady, a senior analyst at Creditsights in New York who spoke to *Forbes* magazine last year about Dow's new CEO, "The new executive will also need to oversee the (continued) legal isolation of Union Carbide. Dow has been very diligent in putting together its structural defense (against Union Carbide's asbestos liabilities) but Dow recently lost a Supreme Court appeal in its West Virginia case."

The question is whether Dow's walls will hold up under more intense legal scrutiny. Dow are gambling that they will: last year they made themselves Carbide's chief creditor in a seeming preparation for sending Carbide into Chapter 11 – a la Dow-Corning at the time of the silicon suit – if asbestos liabilities continue to hit without legislative protection. As chief creditor, Dow would have first rights on Carbide's assets were it to go bankrupt. It's a devious arrangement but there's no guarantee they'll get away with it: number of lawyers chasing Carbide will be looking to rip open the corporate veil.

Source: Tim Edwards, letter sent to the *Boulder Weekly*, 2003.

Elusive extradition (2004)

V. Venkatesan of Frontline *probes the legal basis for U.S. government's rejection of India's request to extradite Warren Anderson, former Chairman of Union Carbide Corporation.*

Warren Anderson

THE yet-to-begin trial in the Bhopal District Court of Warren Anderson, former Chairman of the Union Carbide Corporation (UCC), for culpable homicide in causing the 1984 Bhopal gas disaster received a huge setback when the U.S. government rejected in June the Indian government's request to extradite him. The decision has raised questions about the U.S.' compliance with the letter and spirit of the Indo-U.S.

Extradition Treaty and the Indian government's seriousness in pursuing its extradition request.

Anderson was arrested and released on bail by the Madhya Pradesh police in Bhopal on December 7, 1984, four days after the disaster. Since then, he has been ignoring the Bhopal court's summons. The Central Bureau of Investigation (CBI), which was entrusted with the case on December 9, 1984, filed a charge-sheet against 12 accused, including Anderson, in the Bhopal court on December 1, 1987. The Chief Judicial Magistrate (CJM), Bhopal, issued a non-bailable arrest warrant against Anderson on April 10, 1992, so the extradition proceedings could be initiated.

In October 2002, the CJM issued a fresh arrest warrant against him, after rejecting the CBI's plea to dilute the charge against Anderson from culpable homicide (which would invite imprisonment up to 10 years if convicted) to that of rash and negligent act (up to two years' imprisonment if convicted) in accordance with a similar redress provided by the Supreme Court to an accused, an Indian, in the case.

The CBI, left with no option, forwarded the request to extradite him to the Ministry of External Affairs (MEA). The government, which has all along been avoiding any attempt to seek his extradition under flawed legal advice, finally forwarded the request to the U.S. Departments of State and Justice in May 2003 through the Indian Embassy in Washington. This, after a Parliamentary Committee on Government Assurances indicted the government for the inordinate delay in seeking Anderson's extradition.

On July 2 the MEA conveyed the U.S.' decision to the CBI thus: "The Government of the U.S. has carefully considered the Government of India's extradition request for Warren Anderson and has concluded that the request of the Government of India cannot be executed, as it does not meet the requirements of Articles 2(1) and 9(3) of the Extradition Treaty." E. Ahmed, Minister of State for External Affairs, confirmed this in a written submission to the Lok Sabha on August 18. He denied that the U.S. has refused to extradite Anderson because of his age, as he is more than 80 years old. "Age bears no restriction regarding any person's extradition," he told the House in response to a supplementary question from Rajesh Kumar Manjhi, a member of the Rashtriya Janata Dal. Ahmed also said that the government was currently taking no action regarding Anderson's extradition, which suggests that it has no intention of challenging the U.S. decision.

Article 2(1) of the Treaty says: "An offence shall be an extraditable offence if it is punishable under the laws in both contracting states by deprivation of liberty, including imprisonment, for a period of more than one year or by a more severe penalty."

Article 9(3) says: "A request for extradition of a person who is sought for prosecution shall also be supported by: a) a copy of the warrant or order of arrest, issued by a Judge or other competent authority; b) a copy of the charging document, if any; and c) such information as would justify the committal to trial of person if the offence had been committed in the Requested State."

On September 3, victims of the gas tragedy protested outside a court in Bhopal demanding that Warren Anderson be put on trial in India.

To test the U.S.' claim regarding Article 2(1), it is necessary to know whether the offence Anderson is charged with in India is punishable in the U.S. for a period of more than one year. Experts say that manslaughter – an offence under the U.S. law, which is equivalent to the offence of culpable homicide that Anderson is charged with under the Indian Penal Code – is in fact punishable with more than one year imprisonment in nearly all jurisdictions in the U.S. Manslaughter is defined as the unlawful killing of a human being without malice or premeditation, either express or implied and as distinguished from murder, which requires malicious intent. Although the federal nature of the U.S. legal system means that there are different sentences for manslaughter under the laws of each State as well as a separate sentence provided for under the federal statute, the offence is clearly punishable with more than one year imprisonment under all those laws.

Article 9(3) of the Treaty has three requirements: copy of arrest warrant, copy of charge-sheet, and "such information" as would justify the committal to trial, if the offence was committed in the U.S. One may assume that the Indian government would have had no difficulty in complying with the first two requirements, and would have most certainly annexed copies of the arrest warrant and the charge-sheet with its extradition request. But what "information" would justify Anderson's trial in the U.S., had the offence been committed there?

Article 9(3) deals with both procedural requirements (inclusion of appropriate charging documents, warrant of arrest, and so on) as well as the more substantive question of what is called "probable cause" under the U.S. law. Experts say it is not easy to define the requirement of "probable cause" under the criminal law of the U.S. since the concept has been elaborated and re-elaborated by close to 200 years of jurisprudence. The Black's Law Dictionary defines the concept of probable cause as "an apparent state of facts found to exist upon reasonable inquiry which would induce a reasonably intelligent and prudent person to believe, in a criminal case, that the accused had committed the crime charged." In other words, the term "probable cause" means something more than mere suspicion but less than the quantum of evidence required for conviction on the offence in question.

There is still no basis to suggest that the U.S. has denied the extradition request on the grounds of an absence of probable cause. It may be premature to reach any conclusion about this substantive issue in the absence of disclosure of the exact grounds of denial of extradition. However, experts point out that the Indian government has a fair chance of challenging the U.S. decision in the U.S. Supreme Court, if indeed the State Department based its denial on substantive issues.

The Parliamentary Committee on Government Assurances during the term of the 13th Lok Sabha revealed in its 12th report the opinion of the U.S. law firm, M/s Verner Liipfert, whom the MEA first approached for advice on the feasibility of extraditing Anderson. The firm sought to know whether there were legally mandated safety requirements that were not followed by the Bhopal plant before the disaster. Secondly, it wanted to know who was legally responsible for implementing the safety recommendations made in 1982 by a team of experts of UCC who had conducted a survey of the Bhopal plant. The Indian government provided the firm with detailed answers explaining how the stated safety requirements were not complied with, before the disaster. The government also pointed out that there were reasons to believe that Anderson was personally aware of the safety precautions in place in the UCC plant in West Virginia, and that he did not ensure that safety measures of the same standard were enforced in the plant in Bhopal. In spite of this, the firm concluded that there was no evidence to establish the necessary factual link between Anderson and the gas leak incident in Bhopal.

Soli J. Sorabjee, then Attorney-General, concurred with this view and dissuaded the government from making the extradition request to the U.S. Sorabjee opined that although it was not impossible to furnish the "missing evidentiary links" such as the extent to which Anderson had decision-making control over the safety and design issues of Union Carbide India Limited, and whether he refused to correct the hazard, he was not sanguine about securing such evidence, considering the time and the effort that it would involve. Sorabjee, however, believed that the U.S. would find policy reasons not to surrender Anderson to the Indian government. These are humanitarian concerns such as his age, and the inordinate delay in the Indian government's decision to seek extradition. The U.S. decision shows that these were indeed not the grounds for the rejection of the extradition request.

The U.S. must reveal in detail the specific reasons for rejecting the extradition request if it wants to show that it does not have contempt for the rule of law and the democratic institutions in India. The Indian government, too, must seek specific answers from the U.S., and if necessary, convince it about the merits of its request, in order to avoid the criticism within the country that it did not pursue the case with the diligence it deserved.

Source: http://www.flonnet.com/fl2120/stories/20041008003811000.htm, *Frontline*, India, September 30, 2004.

Does Indian criminal law apply to Carbide and its CEO? (2005)

Ward Morehouse and Chandana Mathur address the central issue in pursuing the criminal charges against Warren Anderson and the Union Carbide Corporation – whether they are subject to criminal law jurisdiction of the Indian courts.

To begin with, the unjust February 1989 settlement should never have quashed the criminal charges. Plea-bargaining under Indian criminal law is not permitted for serious charges such as culpable homicide, even though it is in the United States. For its part, Union Carbide insists that it, in the words of its vice president and General Council, Joseph E. Geoghan, "has never agreed to submit to the criminal law jurisdiction of the Indian courts."

That statement is factually incorrect as an activist raising this issue pointed out to Goeghan and his superior, Robert Kennedy, current Chairman and CEO, at Carbide's annual meeting in 1992. Carbide specifically agreed "to submit to the jurisdiction of the courts of India" in the May 12, 1986, decision of Judge John F. Keenan in the Federal District court in New York, in which Keenan granted Carbide's motion for dismissal of the litigation against it for damages in US courts and its transfer back to India. The reader will note that there was no limitation in that agreement to civil law jurisdiction. Furthermore, Carbide was aware that criminal charges were involved as early as December 7, 1984, when, according to Dan Kurzman, author of *A Killing Wind*, Anderson was presented with a list of the charges being made against Carbide, its Indian subsidiary, and officials of both companies (including Anderson himself).

For his troubles in raising such a question, the activist interlocutor was treated to an indignant lecture by Geoghan, in which he asserted that US corporations and their officials are never subject to the criminal jurisdiction of other countries under "international law" and that the only courts whose criminal jurisdiction he would recognize were US courts.

Goeghan's answer is, of course, wrong on several counts. Reciprocity is a basic principle under international law. If the US government can try Manual Noriega in a criminal court in the United State for actions taken on foreign soil, it is difficult to see why Warren Anderson and Union Carbide (corporations are considered legal persons under US law) cannot be tried in India on criminal charges of culpable homicide. (In Noriega's case, furthermore, the US government arrested him in Panama and brought him forcibly to the United States to stand trial. India thus far has talked only about initiating extradition proceedings to get Warren Anderson to come to India.)

Clarence Dias, president of the International Center for Law and Development and an international human rights lawyer, notes additionally that "the criminal trial restitution is part of the settlement package. You can't take one portion and leave aside all the others."

When Anderson and Union Carbide failed to appear in the Magistrate's Court in Bhopal to answer the charges against them, the court declared them absconders from justice. Warren Anderson has no known assets in India, but Union Carbide Corporation did at the time – namely it's 50.9 per cent holding in its former Indian subsidiary Union Carbide India Ltd. The Chief Magistrate made a valiant effort to seize and harbor those assets as leverage in persuading the corporation to appear in court. By the equivalent of a sleight of hand, Carbide persuaded the Indian Supreme Court to remove those assets from control of the Magistrate's court and put them in the hands of a questionable charitable trust set up in Britain to construct another useless hospital in Bhopal where the medical needs of the survivors can best be met not through an elaborately equipped western-style hospital such as the one that has been built in Bhopal with the Carbide trust money but instead through small-scale community-based undertakings like Sambhavna Clinic. Through a series of maneuvers Union Carbide escaped the reach of the Indian criminal courts.

Source: Ward Morehouse and Chandana Mathur, *Lessons from Bhopal*, Great Barrington, MA: North River Press, forthcoming (2005).

Dow India's argument against presenting Union Carbide to the Bhopal court (2004)

Although the Bhopal court ruled subsequently that by February 15, 2005 Dow must show cause why it should not be asked to make Union Carbide face trial, this application to the court exemplifies the corporate veils and shifting assets that make multinationals so difficult to bring to trial.

BEFORE THE COURT OF THE HON'BLE CHIEF JUDICIAL MAGISTRATE, BHOPAL
CRIMINAL APPLICATION NO – OF 2004
In the matter of Order Dated 15/16th June, 2004

(C.B.I) State Prosecution

V/S

Warren Anderson & Ors. Accused

AND

Bhopal Group for Information and Action

AND

Dow Chemical International Pvt. Ltd.
Having its Office at Corporate Park, Unit No 1
V. N. Purav Marg, Chembur
Mumbai 400071 Applicant (Present)

THE HUMBLE APPLICATION ON BEHALF OF DOW CHEMICAL INTERNATIONAL PVT. LTD. IT IS SUBMITTED AS UNDER:

Notice dated 9th August 2004 along with copy of Application filed by the Bhopal Group for Information and Action (BGIA) Bhopal, above named, have been sought to be served upon the Applicant at its Mumbai office. On perusal of the said Application it is necessary in the interest of justice that certain facts, which are not properly presented before this Hon'ble Court and, perhaps which led to passing of the said Order dated 15/16th June 2004, be placed for consideration before this Hon'ble Court.

At the outset it is submitted that the Dow Chemical International Private, Ltd (DCIPL) was incorporated in India under the provisions of Companies Act, 1956, on 13th February 1998, vide registration No. 11-113551, which is almost 14 years after the Bhopal tragedy. DCIPL is a separate and individual legal entity, which is subsidiary of Dow Chemical Pacific (Singapore) Pte. Ltd. The Singapore Company was incorporated pursuant to the laws of Singapore and in turn is wholly owned subsidiary of The Dow Chemical Company USA (TDCC). Thus it is clear that the DCIPL which is a company situated at Bombay has no direct nexus either in terms of holding or in terms otherwise with TDCC, USA, therefore, it is a wrong statement as has been submitted before this Hon'ble Court can be served upon TDCC through DCIPL. Without prejudice to the aforesaid submissions, it may further be noticed by this Hon'ble Court that Union Carbide Corporation ("UCC") as per the plan of merger was merged in February 2001 with Transition Sub Incorporation, which is a wholly owned subsidiary of TDCC, USA. It is submitted further that UCC even after merger with Transition Sub Incorporation survives as a separate legal entity and continues to be governed by the laws of the State of New York and has a separate corporate existence. The said merger of UCC and Transition Sub Incorporation did not disturb the status and independence of UCC and UCC survived the merger and as stated above even after the said merger UCC continues to be a New York Corporation and has its separate legal entity with its own assets and liabilities. Therefore, it would be wrong to presuppose that DCIPL has any direct nexus with either the UCC and/or TDCC, USA. In this view of the matter, DCIPL will not be able to accept any liability to serve and/or produce the alleged absconder UCC before this Hon'ble Court.

Without prejudice to what is stated here in above it is stated that the above Application filed by BGIA as Intervenor organization cannot and should not be entertained and be acted upon, as the said Applicant, the Intervenor Organization has no locus standi to intervene in the pending proceeding especially when the premier investigating agency of the country, CBI, has investigated the case and is pursuing the prosecution before this Hon'ble Court. It is stated that any attempt on the part of any private person to take the place of the prosecutor and to administer and conduct the trial is against the dictum of law and as such the instant Application by the Intervenor Organization is not maintainable and should be dismissed.

It is stated that the Criminal Procedure Code, 1973 ("Cr.P.C.") confers powers upon a court to try any person facing criminal charges and further mandates that such trial has to be conducted in the presence of the accused person or his pleader in certain cases. The Cr.P.C also contemplates a situation where an accused person fails to submit himself to the jurisdiction of the Court (and provides certain coercive measures culminating in attachment of property of an accused person who may be an absconder from justice). The process provided by the Cr.P.C in respect of compelling appearance of an accused stands completed with the passing of an order of attachment. Thereafter, as and when such accused person appears the trial against such accused commences. The production of a particular accused being declared an absconder is the duty of law enforcing agencies. Cr.P.C does not provide or contemplate any situation where the court after declaring a person as an absconder and attaching his property/assets there after can exercise further powers for securing the custody of such person. In the present case, the Investigating agency has been taking all such steps, since 1984 for securing the presence of the absconding accused, including UCC, which has been declared a proclaimed absconder and whose assets have been attached vide order dated 30th April 1992 of this Hon'ble Court. In view of the above, the Applicant has absolutely no locus to police the police.

It is submitted that DCIPL or TDCC has no connection whatsoever with the Bhopal tragedy and/or prosecution instituted on account of the said tragedy. It may be noted that at no stage of this trial the DCIPL and/or TDCC was ever issued with any show cause notice or was made party to any proceedings either as respondent or Accused.

In view of what is stated herein above, neither DCIPL nor TDCC could be made liable and/or responsible for producing UCC which is a separate legal entity having its own assets and liabilities, nor can they be responsible and/or liable for the alleged criminal acts committed about two decades ago by another distinct juristic person, UCC, as criminal liability, by no stretch of imagination can be inherited/devolved especially after almost twenty years. As stated above UCC survived the merger and continues to be a New York Corporation and is a separate legal entity with its own assets and liabilities. In view of the above, the Application of the Intervener Organization deserves to be dismissed in limine.

Without prejudice to all what is stated above, the Applicants can take recourse to the law of the land in the matter of service to the persons concerned in the litigation directly. The mode of service is clearly laid down and mentioned in the Cr.P.C Chapter VI of the Cr.P.C deals with the process to compel appearance. Section 62 of Criminal Procedure Code reads as under:

"62 – 1 Every summons shall be served by police officer or subject to such Rules as State Government may make in this behalf, by an officer of the court issuing it or other public servant."

"The summons shall, if practicable be served personally on the person summons by delivering or tendering to him one of the duplicate of the Summons."

"Every person on whom summons is so served, if so require by the serving officer, sign a receipt thereof on the back of other duplicate."

It is, therefore, clear that as far as practicable summons which may also include notices in Criminal matters should be served personally on the person. Section 64 creates a situation where if a person summoned cannot, by exercise of *due diligence* (emphasis added) be found, the summons may be served by leaving one of the duplicates with him with some adult male member of his family residing with him. This is an alternative mode of serving summons as stated above which include notice also in criminal matters.

So far as corporate bodies are concerned, the mode of service is incorporated under Section 63 of the Code of Criminal Procedure, 1973, which mandates that the summons on a Corporation such as TDCC may be effected by serving on the Secretary, Local Manager or other principal officer of the Corporation or by a letter sent by a registered post addressed to the Chief Officer of the Corporation, in India. As stated above the present Applicant has no authority to accept summons on behalf of TDCC and therefore, service through the present Applicant is not the correct service and the same would be contrary to the provision of the Cr.P.C.

Further Section 105 of the Cr.P.C. deals with the reciprocal arrangement regarding processes and it, inter alia, states to the effect that in any country or a place outside India in respect of which arrangement have been made by the Central government with the Government of such country or place of service or execution of summons or warrant in relation to criminal matters, it may send such summons or warrant in duplicate in such form, directed to such Court and sent to such authority for transmission as Central Government notification specify. The principle underlying these provisions is that the process of a criminal court must reach to the person concerned to whom it is sought to be served, so that, that person can meet and challenge the allegation made. The present Applicant not being an authorized transmitting authority cannot, with respect, accept task of serving the concerned party.

The Applicant being an Indian Company have respect and regard to the process of this Court, but not being able to accept, though thy have received the Notice of this Hon'ble Court, are herewith returning the notice to the Registry of this Hon'ble Court in terms of what is stated hereinabove.

It is, therefore, prayed that

(A) The Application filed by the intervenor organizations/BGIA be dismissed;

(B) The Order passed by this Hon'ble Court on 15th/16th June, 2004 be recalled as the present Applicant, Dow Chemical International Pvt. Ltd. has no authority to accept the service on behalf of The Dow Chemical Company;

(C) The Notice being directed to be issued to the Applicant, Dow Chemical International Pvt. Ltd. may be discharged;

(D) This Hon'ble Court be pleased to issue any other appropriate order which this Hon'ble Court thinks fit in the present circumstances of the case in the interest of justice;

AND FOR THIS ACT OF JUSTICE AND KINDNESS THE APPLICANT AS IN DUTY BOUND SHALL EVER PRAY.

For Dow Chemical International Pvt. Ltd.
Advocate

Place: Bhopal
Dated: 3 day of September, 2004

Source: Sambhavna Documentation Center, Bhopal.

Bhopal Chief Judicial Magistrate's (CJM) order of 6 January (2005)

This order, requiring Dow Chemical Company, USA to state why it cannot present the absconder UCC in the ongoing criminal case for the '84 disaster, is one of the first legal orders to directly involve Dow in Carbide's liability.

Chief Judicial Magistrate, Bhopal
January 6, 2005
Criminal Case No. (MJC) No. 91 of 1992

State Versus Warren Anderson & others

Bhopal Group for Information and Action, Bhopal, Bhopal Gas Peedit Mahila Udyog Sangathan, Bhopal, and Bhopal Gas Peedit Sangharsh Sahayog Samiti, New Delhi [Assisting the Prosecution]

State Through Not Present

Bhopal Group for Information and Action Through Kishore Joshi, Advocate Present

Bhopal Gas Peedit Mahila Udyog Sangathan Through Abdul Jabbar Present

Dow Chemical International Private Limited Through Sandeep Gupta, Advocate Present

Bhopal Gas Peedit Sangharsh Sahyog Samiti Through Jai Prakash Present

Through this order a decision is being given on the application presented by Bhopal Group for Information and Action on 26.02.04. On 15.06.04 this court had passed an order on the application, that it would not be proper to pass an order on affixing criminal and civil liability on Dow

International Private Limited without listening to the opposite party, and that before passing an order on the application, it would be necessary to hear Dow Chemical International. Keeping these facts in view, this court had issued notice to Dow International, Corporate Park, V. N. Purab Marg, Mumbai, along with a copy of the application to show cause why the application of the applicants should not be granted.

In response to the summons, Advocate Mr. Sandeep Gupta representing Dow India System Private Limited, which merged with Dow Chemical International Limited according to the certificate, appeared before the court, and Dow India Systems Limited and Dow Chemical International Limited, were stated to be the subsidiary company of Dow Chemical International Limited, Singapore. Whereas, the application presented before this court on 26.02.04 mentions that Union Carbide Corporation is a subsidiary company of Dow Chemical Company, Midland, Michigan, USA and it is their application that this company be made a party in the criminal case.

In the light of the above facts and consequent to the order of 15.06.04 a notice be issued through the appropriate legal procedure to Dow Chemical Corporation, Midland, Michigan, USA along with a copy of this application.

Case is fixed for 15.02.05 for reply to the application.

Signed
Anil Kumar Gupta
Chief Judicial Magistrate
Bhopal

Source: Sambhavna Documentation Center, Bhopal - unofficial translation from original in Hindi.

PART 2: TWENTY MORE YEARS

6

Surviving

Continuing crisis

THE COMPENSATION FIASCO is just one aspect of Bhopal's second unprecedented disaster: the complete and tragic mismanagement of the aftermath. This quiet and exhausting catastrophe has been fed and nurtured by the negligence and continuing denial of both corporate and governmental officials, and the unknown and suppressed nature of the consequences of exposure to the gas that leaked. It has been compounded by collusion between the corporation and the Indian government, and the generic effects of class discrimination within the city.

The gas disaster took not only lives but also livelihoods; many affected by the disaster could no longer manage the heavy manual labor they had subsisted by before their exposure. And so Bhopal has become a city of the sick and of those who profit from the sick.

The second section of this book begins with testimonies from the aftermath, and includes the acceptance speeches of survivor-activists Rashida Bee and Champa Devi Shukla, who were awarded the Goldman Prize for Environmental Activism in 2004. The chapter concludes with Suroopa Mukherjee's fascinating piece on the cultural divides of the gas disaster's aftermath within Bhopal.

Voices from the aftermath (1985-1990)

These brief, personal testimonials by survivors illustrate the continuing hardship of the aftermath of the Union Carbide disaster.

Suresh (8) student, class 2, Shakti Nagar

There is very little to eat. Very little to wear. Papa just doesn't get a job. He has no permanent job. Before the leak, he used to work on a boring machine. Now he cannot work on that machine.

Carbide must be punished. Take them to the police station. Then hit them and then jail them – those Carbide fellows. I can't play. I am weak. My hands and legs ache when I run. I get breathless soon. If I run I fall down immediately.

Shammu Khan (50) rents out bicycles, Indira Nagar

People are still going around in circles for their 1500 rupees relief money. The assets of Carbide are still intact. Neither is the government taking it over, nor is it using Carbide's assets to help the poor victims. The people are not quiet, it's just that they are being lulled. Like when a child cries, one soothes it by diverting its attention saying a tiger is coming or a goat is coming. Nether does the tiger come nor does the goat. And the child eventually sleeps. The government is working in a similar fashion. We will have to cry out all over again.

S.K. Dube (32) former UCIL plant operator, Firdous Nagar

At the time of the disaster I was on duty at the Sevin plant. I was also a member of the emergency squad so I stayed on to control the leak. That is how the gas hit me and I became unconscious. I was admitted to the hospital and my lungs were operated upon. Now I face great difficulty in walking. I suffer from breathlessness and the left side of my chest hurts terribly. When I try to read everything appears hazy. I am getting some treatment but it gives only temporary relief. The doctors say there is no treatment for MIC poisoning. The ICMR people call me for tests and they do blood gas analysis, pulmonary function tests urine tests etc. We are never given the reports of these tests and the doctors do not tell us anything properly.

I used to inhale toxic gases even before the gas disaster. All kinds of gases used to leak inside the Carbide plant and the managers never did anything about them. Workers used to be sick quite often – they would vomit, feel giddy and had headaches. And now the gas has made me completely disabled.

Ahmed Ali (45) unemployed, Bapna Colony

Before the gas leak I worked in the textile mill, in the spinning section. When the gas leaked I became very ill, and was admitted in the hospital. I got certificates made there and returned to the textile mill. I told my boss "here is my certificate, give me some lighter work to do." But he said, "We are reducing the number of workers, so we can't give you any lighter work. Hand in your resignation." The company people told me to give it in writing that my health was very bad and that I was resigning of my own accord. If I wrote that, they said, they would accept it, and if not, they wouldn't take me into work anyway. I was a permanent worker, and I had worked there for fifteen years. Now I'm not doing any work. From the day the gas leak till today I haven't done any work. I go here and there in search of work, but no work is available.

Chander Singh (41) auto-rickshaw driver, Karimabaksh Colony

Since the gas leak I suffer from acute breathlessness, my limbs ache and I often get high fever. I have worked as a chef in many of the big hotels in India but now I cannot do this work. When I enter a kitchen where the fires are burning and the spices are being fried I start coughing violently and feel like vomiting. The other thing I could do was to drive, but for two years my hands were numb and I couldn't drive. Now I some how manage to drive an auto-rickshaw and support my family of eight people. Earlier I used to earn three to four thousand rupees a month, now I hardly earn five hundred. I cannot work for more than four to six hours and can only work for 15 to 20 days a month. I have given six applications to the Collector, six to Chief Minister, six to the Commissioner, Gas Relief. I also wrote letters to the Prime Minister and the President about the poor condition of my family. In my letter to the Speaker of the legislative assembly I asked permission to immolate my self along with my family. I was so desperate.

Sunil Kumar (15) student, Jai Prakash Nagar

My brothers Anil and Santosh died from the gas. My parents and three of my sisters – Pushpa, Kiran and Sanju also died in the disaster. Now I live with my sister Mamta and my brother who is three and a half years old. He remains sick very often. Union Carbide has done this to my family. I want to teach them a lesson some day. I am a member of the organisation: "Children Against Carbide" – other children who have lost their parents are also members. The Government has told us nothing about the settlement with Carbide. I read about it in the newspapers and went to the court to find out about it. The case against Carbide must be continued. Till now they have not established who were responsible for the gas leak. The CBI inquiry has also been stopped. The people must know who was responsible for the gas disaster – who killed their loved ones. And those who are found responsible must be hanged. What is the use of all the money if those who have killed so many go scot-free. Similar things will happen elsewhere and there also they will say " take some money and settle the matter." This will become the rule.

Narayani Bai (35), Mahamayee ka Baug

This is the sixth time I have been admitted to the MIC Ward. I have been here since the last month of 1985. When I feel a little better the doctors send me home but I can't stay there for long. My breathlessness become acute and my husband has to bring me back to the hospital. The doctors say that the gases have damaged my lung badly. They say nothing can be done about my disease. Before the gas I had never seen the insides of a hospital. And now I have spent most of the last five years on this hospital bed. I used to work as an assistant at a day care centre and now I can not do any work. My husband Kaluram also cannot go to his job. He used to carry loads. My son works as a tailor, he is the only one earning in the family.

Source: "We Will Never Forget," Bhopal Group for Information and Action, December 1985 – 1990.

Bhopal lives: Sajida Bano's story and Harishankar Magician's (1996)

These two vignettes by Suketu Mehta deal with the difficulties of making a life after the gas. Sajida Bano is a gas widow who had to leave home after the disaster; Harishankar Magician has to adjust his trade to his condition after exposure.

Sajida Bano never had to use a veil until her husband died. He was the first victim of the Carbide plant: In 1981, three years before the night of the gas, Ashraf was working in the factory when a valve malfunctioned and he was splashed with liquid phosgene. He was dead within 72 hours. After that, Sajida was forced to move with her two infant sons to a bad neighborhood, where if she went out without the burkha she was harassed. When she put it on, she felt shapeless, faceless, anonymous: she could be anyone's mother, anyone's sister.

In 1984, Sajiba took a trip to her mother's house in Kanpur, and happened to come back to Bhopal on the night of the gas. Her four-year-old son died in the waiting room of the train station, while his little brother held on to him. Sajiba had passed out while looking for a taxi outside. The factory had killed the second of the three people Sajiba loved most. She is left with her surviving son, now 14, who is sick in body and mind. For a long time, whenever he heard a train whistle, he would run outside, thinking his brother was on that train.

Sajiba Bano asked if I would carry a letter for her to "those Carbide people," whoever they are. She wrote it all in one night, without revision. She wants to eliminate distance, the food chain of activists, journalists, lawyers, and governments between her and the people in Danbury. Here, with her permission, are excerpts that I translated:

Big people like you have snatched the peace and happiness of us poor people. You are living it up in big palaces and mansions. Moving around in cars. Have you ever thought that you have wiped away the marriage marks from our foreheads, emptied our laps of children, bathed us in poison, and we are sobbing, but death doesn't come. Like a living, walking corpse you have left us. At least tell us what our crime was, for which such a punishment has been given. If with the strength of your money you had shot us all at once with bullets, then we wouldn't have to die such miserable sobbing deaths.

You put your hand on your heart and think, if you are a human being: if this happened to you, how would your wife and children feel? Only this one sentence must have caused you pain.

If this vampire Union Carbide factory would be quiet after eating my husband, if heartless people like you would have your eyes opened, then probably I would not have lost my child after the death of my husband. After my husband's death my son would have been my support. But before he could grow you uprooted him. I don't know myself why you have this enmity against me.

Negative-Positive

The gas changed people's lives in ways big and small. Harishankar Magician used to be in the negative-positive business. It was a good business. He would sit on the pavement, hold up a small glass vial, and shout, "Negative to positive!" Then, hollering all the while, he would demonstrate. "It's very easy to put negative on paper. Take this chemical, take any negative, put it on any paper, rub it with this chemical, then put it in the sun for only 10 minutes. This is a process to make a positive from a negative." By this time a crowd would have gathered to watch the miraculous transformation of a plain film negative onto an image on a postcard. In an hour and a half, Harishankar Magician could easily earn 50, 60 rupees ($2) in this business. Then the gas came.

It killed his son and destroyed his lungs and his left leg. In the negative-positive business, he had to sit for hours. He couldn't do that now with his game leg, and he couldn't shout with his withered lungs. So Harishankar Magician looked for another business that didn't require standing and shouting. Now he wanders the city, pushing a bicycle that bears a box with a hand-painted sign: "ASTROLOGY BY ELECTRONIC MINI COMPUTER MACHIN."

Passersby, seeing the mysterious box, gather spontaneously to ask what it is. He invites them to put on the Stethoscope, which is a pair of big padded headphones attached to the Machin. Then the front panel of the Machin comes alive with flashing Disco Lights, rows of red and yellow and green colored bulbs. The Machin, Harishankar Magician tells his customers, monitors their blood pressure, then tells their fortune through the Stethoscope. The fee is two rupees (six cents). Harishankar doesn't like this business; with this, unlike his previous trade, he thinks he is peddling a fraud. Besides, he can only do it for an hour and a half a day, and clears only about 15 rupees (43 cents).

Harishankar Magician is sad. He yearns for the negative-positive business. Once the activist Sathyu took a picture of Harishankar's son, who was born six days before the gas came. He died three years later. Harishankar and his wife have no photographs of their dead boy in their possession, and they ask Sathyu if he can find the negative of the photo he took. Then they will use the small vial of chemical to make a positive of their boy's negative, with only 10 minutes of sunlight.

Source: Suketu Mehta, *Village Voice*, December 3 and December 10, 1996.

The Goldman Foundation Environment Award (2004)

This speech was delivered by prize winners and Bhopal activists Rashida Bee and Champa Devi Shukla when they received their "Green Nobels." Bi and Shukla, both survivors, have been running a trade union of gas exposed women workers in an office stationary production unit for eighteen years.

Rashida Bee

On behalf of the people of Bhopal and ourselves, we greet our brothers and sisters here. When I learned that sister Champa and I had won this huge award our first response was that of a long silence. We knew a few individuals who had won awards. They were all educated people, spoke English and had email accounts. Has there been a mix up? We wondered.

As you have just seen in the video, we – the victims of the world's worst industrial disaster – are fighting against the world's number-one chemical corporation. We know that Dow and other such corporations are responsible for slow and silent Bhopals all over the world. We know that not just in Bhopal, but mothers everywhere in the world, carry chemical poisons in their breasts. Bhopal is but the most visible example of corporate crime against humanity.

We are aware that the day we succeed in holding Dow Chemical liable for the continuing disaster in Bhopal, it will be good news for ordinary people all over the world. From that day forward, chemical corporations will think twice before producing and peddling poisons and putting profits before the lives and health of people.

We are not expendable. We are not flowers to be offered at the altar of profit and power. We are dancing flames committed to conquering darkness. We are challenging those who threaten the survival of the planet and the magic and mystery of life. Through our struggle, through our refusal to be victims, we have become survivors. And we are on our way to becoming victors.

Champa Devi Shukla

When sister Rashida and I got the news of the award, faces of friends and comrades in the struggle for justice in Bhopal swam before our eyes. In our hearts we include them all as co-recipients of this great honor. Once again, we would like to acknowledge and honor the contribution of our sisters and brothers in Bhopal and other parts of the world who have helped us become more than what we are.

Sister Rashida and I have decided that we will hand over the entire sum of the award money to a trust that will provide medical assistance to children born with deformities, run income-generating projects for women survivors and institute an award for ordinary people fighting extraordinary battles against corporate crime in our country.

Source: Rashida Bee and Champa Devi Shukla, Goldman Website, April 19, 2004.

Anger and denial on the streets of Bhopal (2004)

Suroopa Mukherjee looks into the conflicts within Bhopal, between the sick and struggling of Old Bhopal, different activist groups, the decision-makers and bureaucrats of New Bhopal, and the blame game that keeps them all apart.

On the night of December 2/3, 1984, almost 40 tons of methyl isocyanate (MIC) spewed out of a storage tank at the Union Carbide India Ltd (UCIL) factory in Bhopal.

Who were the real victims of this disaster? Newspaper clippings tell us they belonged to the lowest economic strata, people living on the fringes of society. Was this accidental? No, for the line of separation between the two parts of the city is marked by the Upper lake. Its shimmering waters absorbed the gas and prevented MIC from spreading to New Bhopal, where the rich and more privileged members of society reside.

Those who lived in Old Bhopal breathed in the deadly concoction of gases that caused multi-systemic injuries. Today their lungs, eyes, reproductive systems, immunological levels and mental health are severely impaired.

How did the gas affect the rest of the city? A relative with whom I stayed, in Arera Colony, the first time I visited Bhopal described the night quite grimly. "At first, we were not even aware of the event. But then a doctor friend rang up. We took our car and went some distance and saw people straggling along in groups and trying to run. We heard people shouting and crying. The magnitude of what happened became clear in the morning when we heard the news. It must have been very bad. We have heard the extent of suffering from friends who worked for the victims. But Bhopal has also become a media story. It is so difficult to separate fact from fiction."

How, indeed, does one separate the fact from the fiction? One way is to explore the attitudes of those who were not directly affected, but whose opinions determine how Bhopal is popularly perceived. This includes the decision-makers of New Bhopal: people in government, whose point of view feeds into the policies made and implemented by the State. Like the Director of the Bhopal Cell, Ministry of Gas Relief and Rehabilitation, who said rather bitterly, "Our office is being asked to look after these people for the rest of their lives and now the next generation as well. When will it all stop? Where is the money and the resources? How long will Bhopal pay the price for a single event that was essentially an accident?"

What makes the Hiroshima of chemical disasters a continuous tragedy is the way in which it has been written off as a "one-time event" whose aftermath was contained by a monetary settlement.

In Bhopal itself, everyone is trying to forget the world's worst industrial disaster. But they are compelled to remember. This is where the conflict is. This is perhaps why there is so much anger and denial everywhere in Bhopal. Doctors, bureaucrats, intellectuals, students, traders, ordinary citizens respond with hostility to any questions pertaining to the disaster. To hear them speak is to understand the invisible line that distinguishes fact from fiction. Denial is as much a mental affliction, and the violence and suffering it unleashes is evident on the streets of Bhopal, even after 20 years.

In management parlance, the best way to classify an event is to identify the stakeholders and see their response. The pattern that emerges is fairly typical, and it helps to arrive at neat and conclusive definitions. But Bhopal has a strange way of evading definition. We begin with doctors and bureaucrats who were called upon to take "charge" of the disaster hours after it occurred. Why should the truth be elusive in their case?

Dr. N. P. Mishra, who was dean of Gandhi Medical College in 1984, describes the night as overwhelming and unprecedented. "The first problem was that of numbers. My team treated 170,000 patients in one day. The second problem was lack of information. UCC continuously informed us that the gas was not toxic and that we should not apprehend long-term effects of the gas. They insisted that most of the casualties were a result of panic and people running. But I took it in my hands to organize bulk supplies of medicine. I rang up colleagues and civil surgeon friends in neighboring towns like Indore, Hoshangabad and Vidisha and asked them to send supplies of medicine and necessary staff like nurses and ward boys. I called up local chemists and asked them to pool in their stocks. Payment, I said, would be made later. Subsequently, I was accused of taking steps without prior government permission. I find this kind of attitude unfortunate. Seventeen thousand nine hundred patients were admitted and treated in an 850-bed hospital. Tell me honestly, can the role of doctors be underplayed in the Bhopal gas tragedy?"

Dr Mishra is far more hesitant when I ask him what kind of research was initiated to study the impact of MIC on the human body. "I am afraid research had to take a back seat because we were too busy managing patients. Four-five days later, the Indian Council of Medical Research (ICMR) stepped in and 27 projects were set in motion. The then DG told us that money was not a constraint. I realize now that this was a research opportunity of an unprecedented kind. But there was bureaucratic intervention at every step. There was a ban on the publication of data, and we had to take permission from the central government to attend seminars, speak at any public forum or to the media. And then, just as suddenly, in 1994 the projects were terminated so that many of the findings remained inconclusive. I was appalled to read a press statement made by the present DG that most of the research done here in Bhopal was sketchy. I vehemently deny it. It was excellent work but the findings were suppressed."

I am sitting in Dr Mishra's room and I can see the walls are decorated with awards, citations and pictures taken with dignitaries. Outside is his OPD, and patients are waiting in large numbers. He tells me that even today he sees gas patients, and though he cannot treat them free and much of his diagnosis is symptomatic, in the absence of significant findings, he works very hard to bring relief to them. When I ask him whether patients continue to suffer the after-effects of the gas, that UCC was proved wrong, his answer is evasive and echoes what most other doctors say: "Do not be taken in by the big rush at gas *rahat* (relief) hospitals. There is no way of knowing the true from the false cases. You see, each one of them is given a card that entitles them to free treatment. So families flock to the hospitals and hordes of relatives jump onto the bandwagon."

When I ask doctors why there is so much anxiety about "true" or "false" cases, and why it is used as a parameter to define a gas victim when no such distinction can be part of any treatment protocol, they reply defensively. At the same time, this criterion becomes the basis for arriving at conclusions with far-reaching implications. Dr Rachna Pandey, who was then a post-graduate student at Hamidia Hospital and was working on the clinical medicine project, is categorical when she says that no research is possible in Bhopal today. She admits that if the "projects had continued, path-breaking facts would have emerged. But now all one can have are speculations and postulations. Real and fake facts have got so mixed up that no retrieval of original data is possible. Too much time has been lost. Maybe that is the real tragedy."

As I listen to her I realize what is happening in Bhopal today. Attempts are being made to put together parts of a jigsaw puzzle, but the picture that emerges is so smudged that it cannot appear real from any angle. There is large-scale generalizing. I am repeatedly told: "Slums in any part of India are no different from the gas-affected *bastis.* Who drinks clean water in India? Are not most women in India anaemic? Are we not dealing with large-scale poverty and destitution and beggary? There are toxins in everything we eat. If you say that there are more cancer cases in Bhopal then maybe Bhopalis smoke more. Given the statistics for TB, cancer or any other disease, how different is it from the rest of India?"

The biggest bugbear in this disaster is the compensation money that was paid to survivors. How does it matter that it is a pittance? Or that it will be used only to pay hefty medical bills, or debts? Post-1984, the widespread ailment that grips Bhopal can indeed be termed "compensation neurosis." But it is something that has affected everybody. Almost everyone I spoke to blamed the government for not working out a parameter to separate the deserving from the undeserving. Doctors and bureaucrats, for all their mutual suspicion, agreed that money had corrupted the Bhopal survivors. Collectively they said: "A large section of this population has become lazy and greedy. We cannot get domestic help in Bhopal because that section of society lives on the dole. NGOs and activists back them. They take to the streets and voice their complaints. Every gas victim in Bhopal is a politician who takes advantage of a corrupt system."

The bureaucrats I meet are vociferous in the blame game. Iqbal Ahmed, principal secretary in the ministry of relief and rehabilitation, is candid when he says that the state is up against the Centre. "My coffers are empty. It is easy for the Supreme Court to give orders to supply water through pipelines and tankers. But where are the funds? No, you are wrong to say that we have asked for money from the Rs.1,500 crore surplus lying with the Reserve Bank. We have simply said give us the money from wherever you want. We do not want the victims to pay, but then who pays? What about the mammoth task of cleaning up the contaminated site? Do we have the money and the technology? It is easy for people to make noise in the press. I am glad that now the US court has said that Union Carbide will have to do the cleanup. We have asked an Indian agency to look into the modalities. We are working day and night but the task is not easy. I cannot answer your question about why Madhya Pradesh took back the land. Then you go back to the beginning and ask why the Supreme Court made a settlement. What was the government at the Centre doing then? The questions are endless and I do not have the answers."

The director of the Bhopal cell, Bhupal Singh, is more ready with his answers. "What is the point of giving these people more money? They will spend it on buying consumer durables like colour televisions and what not. Soon Bhopal will be flush with surplus money and this will lead to inflation. So all we said was let the money be kept in a central kitty, and let it be used for long-term treatment. Please understand: the tragedy has been a disaster for some and a boon for others." I am reminded of what the former chief secretary to the government of Madhya Pradesh said to me at an earlier meeting: "The problem with the Bhopal survivors is that they want the government to hold their hand forever. I agree that earning capacity has gone down with physical disabilities and that lighter options should be provided, but you have to be rational. Grants and loans are well within the capacity of the government and they are available. Look, even if bureaucrats are willing, politicians do not listen. Vested interests and corruption are rampant at every level, and where is the system for accountability?"

It is surprising that almost all the official stakeholders are reluctant to say that a national commission on Bhopal, comprising eminent people from different walks of life, should be set up to constantly monitor the situation. Some react angrily: "Is it not bad enough that the Supreme Court is monitoring us all the time? And orders are simply passed to do this or that without going into the ground realities. Besides, any national commission will soon become bureaucratic and be mired in red tape. Mark my words, everybody passes the buck in this country." The former chief secretary is more amenable: "OK, a scientific body, maybe."

Why is it that the scientific approach with its emphasis on objectivity is a failure in Bhopal, I ask myself. But then, how can denial and objectivity go hand in hand? The virulence that one faces in response to questions can best be summed up by a statement made by the chief medical officer in the presence of the director and almost all the superintendents of gas hospitals. "The truth of the matter is that there is no way of knowing scientifically whether people are still suffering the after-effects of the gas. In other words, 95% of people in Bhopal are *not* gas victims." Heads nod in unison and a superintendent butts in: "The fact is that if we had been honest in defining the number of 'real' gas victims the number would not have crossed 30,000. But we said: 'Let it be, these are poor people, if by identifying them as gas victims they will get benefits, what harm?'. We have been more charitable than negligent." To my volley of questions about why have a gas relief and rehabilitation ministry if that be the case; and on whom is the allocated budget being spent if the percentile is only 5%; and why is the state government refusing to come out with a white paper on expenditure, the replies are angry, with counter-accusations of unwarranted and unethical questions. Denial and brutality go hand in hand, and I realize that I am as angry and frustrated at the end of the day. Imperceptibly, the gas has begun to affect all our lives.

The other two stakeholders are the political activists and the survivors. The activists are wide-ranging. Some claim to be gas-affected Bhopalis with distinct political affiliations, others are middle class and academic – people who stayed behind in Bhopal and who believe in combining information and action. Some are women and youth leaders. The groups are fractured and not always in agreement about ideological positions and modes of operation. In a sense they too are sucked into the wilderness of vested interests, and the corridors of power. There is much talk in Bhopal about NGOs making money. As one official put it: "People claim to be working for the downtrodden but they are full of themselves, seeking media attention and credit for what is done. How much work do they really do? Yes, there are some who work but they are unsung and unknown."

Meeting individual leaders is to encounter a closed exterior that reveals nothing of the inner turmoil. They nourish their own constituencies and work for the benefit of their own groups. Many of them have loyal followers and I am taken to meet them.

The gas survivor is perhaps the most ambivalent of stakeholders. The physical and mental scars are palpable, so is the anger and frustration with the system. They do not know whom to blame. They view me with a mixture of suspicion and trust. The fact that I have come with a group of students who represent the youth and future of the country makes them curious. The fact that we are from Delhi , the seat of power, makes our role crucial. They flock to us and ask for details to be conveyed to the *sarkar*, as they describe the government at the Centre. Their demands are chillingly simple: clean water, proper medication and means of livelihood. But what they are really asking for is far more complex: "Give us justice not money, give us back our lives and, please, listen to us," is what they say repeatedly.

When I take the students to Neelam Park where a survivor group has been holding meetings every Thursday and Sunday for the last 20 years, we meet a group of widows. Once the initial introductions are over, and the hesitancy dispelled, they hold our hands and cry. Many recount their stories with the usual compulsion that marks narration in Bhopal. How does one distil fact from fiction? I watch the rows and rows of faces, gnarled, wasted, sick, full of despair and hope. When one of the students asks them why they come here every week, a woman who identifies herself as a group leader of sorts says: "We will continue to fight for justice till the day we die. We lost everything on that night, we have nothing else to lose, so we have no fear." I realize that activism in Bhopal has percolated down from the leaders to the nameless faces, and individual constituencies nurtured by different political groups cannot bury the similarities and the differences. I see that many of the women are burkha-clad, their ages ranging from the very young (second-generation, born to gas-affected parents) to the very old. They carry papers and documents as proof of their identity, and when that fails they reveal the scars on their bodies. They are not hesitant to face the camera or to talk to us.

In New Bhopal the faces turn away from the camera and people refuse to comment. To them Bhopal is no longer a part of their living reality. They are too busy making their own careers and earning their living. As one young woman retorted: "I am a student of architecture. I have no time for social work." To others, our motive as a group is suspect. What are we doing here – research, investigation or social activism? "People like you come and go but nothing happens," they say bitterly.

Every night we have our group meetings where each of us recounts the day's happenings. Our own narratives have become repetitive and strangely inconclusive, but our energy levels have not dissipated. Bhopal makes us speak out and protest. We are largely students of literature (and from most counts fiction is our area of interest), but here we are grappling with a different discourse, language and meaning. For many of us this is our first experience of fieldwork that takes us outside the closed academic circle. As we exchange notes we comment that we find management's use of the stakeholder model inadequate. What we are talking about is a man-made disaster that affected lives in different ways. Twenty years have passed and the scars remain on the psyche, hidden or palpable. Any kind of classification only creates artificial distinctions between true and false, and separates the beneficiaries from the losers. In the process, the pressing issues of justice, liability and responsibility get blurred. As one student succinctly puts it: "I have learnt more about systems of oppression in these few days than I have from all my readings. What is staring us in the face is class and gender discrimination, based on how power operates. Bhopal is a modern-day tragedy that had to happen one way or the other."

What we intend to do with our "fact-finding" is to write a report accompanied by a video presentation. We then plan to take it to all the officials who hold positions of power in Delhi . We want to tell them why and how Bhopal continues to matter even today. We also want the collective strength of youth, that We for Bhopal as a group represents, to be felt in the right quarters. If in the process we are able to counter the denial that is part of the popular perception of Bhopal then we would have made a small but significant beginning.

(Suroopa Mukherjee is staff advisor of We for Bhopal, a students group based at Delhi University and committed to the fight for justice for survivors of the Bhopal gas tragedy. The aim is to create awareness about the pressing issues of environmental pollution, violations of human rights and corporate crime)

Source: InfoChange News & Features, December 2004, www.infochangeindia.org.

7

Sickness

Endless ills

SURVIVORS HAVE SAID that the fortunate ones were those who died immediately: their suffering was over. Those who did live have faced not only illness but continuing uncertainty about their health, their future, and their children. Women don't know whether they should give birth; whether their bodies can produce healthy life. Doctors don't know how to treat the gas affected.

On the night of the disaster the Union Carbide doctors told the panicked Bhopal hospital staff that MIC was no worse than a tear gas. Wash the eyes with water, they said – it will pass. Although, as the limited medical studies excerpted in this chapter show, the effects of the gas are clearly systemic and in many cases permanent, Union Carbide has provided no further information about the chemical they created and used. However, they have done research. For example, in their manual entitled Methyl isocyanate F-41443A-7/76, they cite a paper "U C Pozzani and E R Kinkead, *Animal and Human response to Methyl Isocyanate*, a paper presented at the May 16-20, 1966 Meeting of the American Industrial Hygiene Association, Pittsburgh, Pa." This paper has never been released. The same manual cites in addition "Unpublished work of Union Carbide Corporation." Union Carbide officials have in fact done their best to provide disinformation about the gas leak. By down playing the effects of MIC and withholding medical information about MIC, they attempt to lessen their own liability.

An unprecedented mass exposure to MIC would be difficult to address under any circumstances; but in Bhopal the attempts to heal the worst affected have been additionally hindered by preexisting poverty and enduring class discrimination. Economic hierarchies and communal differences were, and remain, firmly entrenched in the city of Bhopal. The doctors and bureaucrats, drawn from the upper echelons of society, often discriminate openly against the lower classes. And because, lacking statistical evidence and a diagnostic protocol, the symptoms of poverty and the symptoms of gas exposure can resemble each other, many of those entrusted with curing the gas affected would rather just diagnose, and ignore, the poverty itself. Sickness, impeding work, causes increased poverty and increased sickness, creating a miserable and unbroken cycle.

This dynamic is reinforced and enabled by the non-visible nature of the illness caused by the gas, and similarly the water contamination. The leaking gas did not blister the skin; its symptoms are multifarious and often invisible; they become identifiable and distinguishable only by carefully and scientifically documenting the health of the affected communities across generations. As the studies excerpted in this chapter illustrate, the gas-affected communities do continue to get sick and to die at much higher rates than normal, and to suffer from particular symptoms – some of which defy the available vocabulary. However, as nearly all of these studies conclude, the research needed to develop an understanding of the effects of the gas, and to intelligently treat them, is still grievously lacking.

This dearth of research is itself criminal. Immediately after the worst industrial disaster in history, the Indian Council for Medical Research (ICMR) began eighteen long-term, large-scale, government funded medical studies of the first and second-generation effects of gas exposure. These studies, the only ones that had the funding, timing, and scope to understand the impact of the disaster, were suddenly and prematurely terminated in 1994 with no explanation. When their limited conclusions were finally released in 1996, the scientists were only able to ruefully note that "It would be desirable to extend the long-term study of the same cohort in Bhopal." Some suggest that the studies were terminated when conclusive evidence of second-generation damage began to emerge, an eventuality that would have radically increased the liability of the corporation, and perhaps the government.

Independent scientific and medical studies, such as the report of the International Medical Commission on Bhopal (IMCB), excerpted in this chapter, or the study on second generation effects published recently in the *Journal of the American Medical Association* (JAMA), have unfortunately been unable to replace the government studies due to their sporadic nature and limited funding. Additional work, such as Daya Varma's study on the effects of MIC on pregnancy, and Ramana Dhara's report on the lack of information about the gas, underline the desperate need for further research on the health consequences of the Bhopal disaster. Meanwhile, in Bhopal, sickness is a daily reality.

Epidemiological and experimental studies on the effects of methyl isocyanate on the course of pregnancy (1987)

Daya R. Varma, one of the few scientists to study the effects of MIC on living systems, conducted research on its relationship to reproduction, and its effect on pregnancy through tests on mice.

Abstract

Although press reports indicate that the leakage of methyl isocyanate (MIC) on December 3, 1984, in Bhopal has led to an increase in spontaneous abortions, stillbirths, infant mortality, and fetal abnormalities no clinical or experimental studies on the reproductive toxicity of MIC were reported in scientific journals for several months after the accident. We therefore conducted, 9 months after the accident, a preliminary survey of 3270 families in Bhopal and experimental studies on the effects of MIC in pregnant mice. It was found that 43% of pregnancies in women residing near the Union Carbide pesticide plant did not result in the birth of a live child. Likewise, exposure of mice to relatively low concentrations of MIC (9 and 15) for 3 hrs caused complete reabsorption in more than 73% of animals. A decrease in fetal and placental weight was observed at 2 to 15 ppm MIC. In general, the experimental findings in mice corroborate the epidemiological data from Bhopal. The mechanism of the fetal toxicity of MIC remains to be established.*

Discussion

The epidemiological and experimental data reported here clearly indicate that the exposure to methyl isocyanate can adversely affect the course of pregnancy. These studies confirm the reported increase in stillbirths and spontaneous abortions in Bhopal in the post accident period (2.5). However, a more controlled and systematic epidemiological study than the one reported in this paper and a longer follow-up are needed to more precisely establish the magnitude and nature of the adverse effects of the MIC leakage on the course of pregnancy in Bhopal.

Although the survey relied entirely on the information provided by the victim or a senior member of the family and does not involve any objective assessment, the data are probably as reliable as they can be. For example, there is only a 2% discrepancy in relation to the reporting of a pregnancy terminated by abortion, as compared with the reporting of a pregnancy not culminating in the birth of live child. The higher incidence of pregnancy loss in women who were in their first trimester relative to those who were in their second or third trimester at the time of the accident could have been recorded if women mistook delayed menstruation or excessive vaginal bleeding as abortions. Because of the time gap, the survey could not ascertain the incidence of abortions in the 2 years preceding the Bhopal methyl isocyanate incident. Nevertheless, a 43% pregnancy loss in the post accident period is three to four times higher than the normal incidence (as estimated by the Indian Council of Medical Research) of 6 to 10% abortions in Bhopal (5) Also the rate of infant mortality within 30 days after birth is five to six times higher than that recorded for 1984 (2 years preceding the accident).

In many ways, the experimental data confirm the clinical findings. For example, at 9 and 15 ppm MIC, more than 75% of mice lost all of their fetuses; however in animals that did retain their pregnancy, the litter size was not dramatically reduced. This is unusual because with most agents that cause fetal toxicity, there is a noticeable increase in the loss of implantations, although most animals tend to retain some viable fetuses.

From this study it is not clear if the fetal toxicity is due only to effects on the mother or due to both maternal and fetal effects. The greater body weight loss in the first 2 days after the exposure to MIC in pregnant animals that did not retain pregnancy as compared with those that did retain pregnancy indicates either an early onset of fetal toxicity or a relationship between marked maternal toxicity and ultimate fetal loss. The magnitude of hypoxia that result from these exposures (8,12) may not explain such marked fetal toxicity in light of the fact that in mice, marked hypoxia induced by other techniques was found to exert far less fetal toxicity than was observed in these other studies (13).

References

1. Anonymous. "Calamity at Bhopal," *Lancet* ii. 1378–1379 (1984)
2. Varma, D. R. "Anatomy of methyl isocyanate spill in Bhopal" in: *Hazard assessment of Chemicals*, Bol.5 (j. Saxena, Ed.) Hemisphere Publishing, New York, 1987, pp.233–289.
3. Bang, R. "Effects of the Bhopal Disaster on the womens' health – An epidemic of gynecological disorders." In: "Selected Documents on the December 1984 Union Carbide Chemical Incident," V. Pinto and S.D. O'Leary, Eds, Washington Research Institute, San Francisco. C.A. 1985 pp.35–37,
4. Shilotri, N. Raval, M.Y. and Hinduja, I.N. "Report on gynecological examination" In: "Medical Survey on Bhopal Gas Victims between 104 Days after Exposure to MIC." Gas Nagrik Rahat Aur Punarvas Samiti, Kohefiza, Bhopal, 1985, pp.16–17.
5. Lepkowshki W. "Special Report – Bhopal." Chem. Eng. News 63 (48):18-32 (1985).
6. Deve, J.M. "The Bhopal methyl isocyanate (MIC) incident: An overview." In: *Proceedings of an International Symposium on Highly Toxic Chemicals: Detection and Protection Methods* (H.B. Schiefer, Ed) Saskatoon, University of Saskatchewan, Canada, Sept. 1985, pp.1–37.
7. Varma, D.R. Ferguson J.S. and Alarie, Y. "Reproductive toxicity of methyl isocyanate (MIC) in mice." J. Toxicol. Environ. Health. In press.
8. Ferguson, J.S. Schaper, M. Stock, M.F. Weyel, D.A. and Alarie, Y. "Sensory and pulmonary irritation with exposure to methyl isocyanate." Toxicol Appl. Pharmacol. 82: 329–335 (1986).
9. Wilson J.G. "Methods for administering agents and detecting malformations in experimental animals." In: *Teratology: Principles and Techniques* (J.G. Wilson and J. Warkany, Eds) University of Chicago Press, Chicago 1965, pp.262–277.
10. Dawson A.B. "Note on staining of skeleton of cleared specimens with alizarin red." S. Stain Technol. 1:123–124 (1926).
11. Mulay, S. Varma, D.R., and Solomon, S. "Influency of protein deficiency in rats on hormonal status and cytoplasmic glucocortionod receptors in maternal and fetal tissues." J. Endocrinol. 95: 49–58 (1982) .
12. Nemery, B. Dinsdale, D. Sparrow, S. and Ray D. "Effects of methyl isocyanate on the respiratory tract of rats." Br. J. Ind. Med 42; 799–805 (1985).
13. Ingalls, T. H. Curley, F.J. and Prindle, R.A. "Experimental production of congenital anomalies." N. Engl. Med. 257; 758–768 (1952).

Source: *Environmental Health Perspectives* Vol. 72, pp.153–157, 1987. *Experimental studies on mice were conducted at the Graduate School of Public Health, University of Pittsburgh, Pittsburgh, PA.

Clinical dilemmas (1995)

This selection from a paper by Ramana and Rosaline Dhara explores the dilemmas facing those striving to provide effective medical treatment to the gas disaster survivors.

Following the Bhopal gas release, it became evident that a large information vacuum about MIC existed not only in India but also globally. The scale of the disaster, the lack of information about what toxicants had been released, as well as the paucity of information about the toxicology of MIC "as an intermediate product" had created much confusion and scientific debate about the medical management of the gas-affected victims. The medical system in Bhopal was further tested by the twin factors of immediate death and serious morbidity of many people and the absence of a definite protocol for the treatment of the poisoning. Although the major hospitals in the area adopted standard protocols to treat the ocular and respiratory effects of MIC, most patients were treated on a symptomatic basis. Local medical practitioners experimented with a wide range of drugs and therapeutic interventions in attempting to provide relief to the gas-affected population.

The questions of clinical and toxicological importance that emerged from the episode, and that persist to this day, refer mainly to the bioavailability of the toxicants, the persistence of multisystemic health effects, and appropriate methods of treatment.

Was MIC bioavailable? Studies using radiolabeled MIC have shown that the compound is absorbed and distributed systemically in animals. These studies provide a possible physiopathologic basis for the occurrence of multisystemic symptoms in the Bhopal survivors.[30-32] Persistent gastrointestinal and neurobehavioral problems have been reported by the Bhopal survivors. Neuromuscular symptoms in the Bhopal survivors have persisted since the gas leaked and include tingling, numbness, a sensation of pins and needles in the extremities, and muscle aches.[33] Neurobehavioral tests conducted two and a half months after the accident showed that auditory and visual memory, attention response speed, and vigilance were significantly impaired in a group of 350 exposed subjects as compared with controls. No effect on manual dexterity was observed.[34] Following a recent investigation of the long-term health effects of the disaster, the International Medical Commission on Bhopal recommended that the spectrum of disease attributed to the disaster be broadened to include neurotoxicity and post-traumatic stress disorders.[35]

Concerns that the gas leak had effects on reproductive health were raised early in 1985, when reports indicated that menstrual cycle disruption, leukorrhea, and dysmenorrhea had occurred in gas-exposed women.[36] The fetus was considered to be at risk not only because of exposure to the toxicants but also because of other factors such as stress and anoxia. An epidemiologic survey by Varma[37] observed increases in pregnancy losses and infant mortality in gas-exposed women. Animal experiments conducted by Schwetz *et al.*[38] and Varma *et al.*,[39] exposing pregnant mice to MIC by the inhalation route, demonstrated a fetotoxic effect. In a further follow-up, Varma *et al.*[40] studied the contribution of maternal hormonal changes and pulmonary damage and showed that the fetal toxicity of MIC was partly independent of maternal pulmonary damage and that MIC could be directly fetotoxic. In the case of Bhopal, the results of the animal studies considered in conjunction with the findings from human epidemiology suggest that exposure to MIC is fetotoxic and that this probably is the result of a direct effect on the fetus.

Other issues that compounded medical management were the circumstances leading to the gas release (hydrolytic and exothermic reaction of water with MIC), raising the possibility that impurities (phosgene) or decomposition products (hydrogen cyanide, nitrogen oxides, and carbon monoxide) were present in the gas cloud. Were there toxicants other than MIC released in the accident? Although studies undertaken by the ICMR[33] to determine chemical composition of the MIC tank residues may shed some light on the toxicants released, many of the questions may be answered only by recreation of the actual conditions leading to the accident.

End notes and References

30 Dhara V.R. "On the bioavailability of methyl isocyanate in the Bhopal gas leak" (editorial). Arch Environ Health. 1992; 47:385–6.

31 Ferguson J. S, Kennedy A. L, Stock M. F, Brown W. E, Alarie Y. "Uptake and distribution of 14C during and following exposure to [14C] methyl isocyanate." Toxicol Appl Pharmacol. 1988; 94:107–17.

32 Bhattacharya B. K, Sharma S. K, Jaiwal D. K. "In vivo binding of [14C] methyl isocyanate to various tissue proteins." Biochem Pharmacol. 1988; 37:2489–93.

33 Indian Council of Medical Research. Bhopal Gas Disaster Research Centre. *Annual Report*. Bhopal, India, 1990.

34 Gupta B. N, Rastogi S. K, Chandra H, *et al.* "Effect of exposure to toxic gas on the population of Bhopal: Part I – epidemiological, clinical, radiological and behavioral studies." Ind J Exp Biol. 1988; 26:149–60.

35 International Medical Commission on Bhopal. "Press release," January 24,1994, New Delhi, India.

36 Bang R, Sadgopal M. "Effect of Bhopal disaster on Women's health – an epidemic of gynecological diseases" (Part I), Unpublished data, 1985.

37 Varma D. R. "Epidemiological and experimental studies on the effects of methyl isocyanate on the course of pregnancy." Environ Health Perspect. 1987; 72:151–5.

38 Schwetz B. A, Adkins B. Jr, Harris M, Roorman M, Sloan R. "Methyl isocyanate: reproductive and developmental toxicology studies in Swiss mice." Environ Health Perspect. 1987; 72:147–50.

39 Varma D. R, Ferguson J, Alarie Y. "Reproductive toxicity of methyl isocyanate in mice." J Toxicol Environ Health. 1987; 21:265–75.

40 Varma D. R, Guest I, Smith S, Mulay S. "Dissociation between maternal and fetal toxicity of methyl isocyanate in mice and rats." J Toxicol Environ Health. 1990; 30:1–14.

Source: Rosaline Dhara and Ramana Dhara, "Bhopal – A Case Study of International Disaster," *International Journal of Occupational and Environmental Health*, January/March 1995.

Report of the International Medical Commission on Bhopal (1994)

A summary of the report of the International Medical Commission on Bhopal (IMCB) by one of its members, Birger Heinzow, MD.

Background:

Following the gas leak from the Union Carbide plant on 2-3 December, 1984, in Bhopal India, thousands of people died and hundreds of thousands were injured. The medical system was overwhelmed in the acute phase of the disaster. The immediate emergency response appeared to be swift, appropriate and effective under the given circumstances.

Some studies (many so far unpublished) have been performed after the disaster, many of them observational and uncontrolled. Most disturbing is the absence of a systematic and reviewed medical strategy for assessing injury, long-term medical care, rehabilitation, fair compensation and coordinated research. The purpose of the IMCB mission was of a humanitarian and medical nature, mainly to contribute to the relief of the victims. The Commission did not intend to provide direct medical care to individuals, which remains the responsibility of the government of India. IMCB operated in Bhopal from 10–12 January 1994, in close cooperation with Indian health professionals, victim and activist organizations.

The idea for IMCB originated directly from the work and the recommendations of the Permanent Peoples' Tribunal (PPT) 19–24 October 1992. An International Commission was then recommended to provide an independent, in-depth assessment of what happened to the victims of Bhopal over ten years post-disaster, and of what occurred which prohibited and delayed appropriate health care delivery. The establishment of IMCB was seen as a first step in the direction of a more direct involvement of the international community. The main task was to investigate the present situation and provide possible recommendations relating to diagnostic methodology and the treatment of the "Bhopal-illness," clinical organization and delivery of medical services, rehabilitation, compensation and social welfare.

Operation of the IMCB:

Between October 1993 and January 1994, members of IMCB were chosen on the basis of needed expertise, from a list of a volunteer medical professionals, and they began corresponding with one another, developing protocol and investigating plans. On January 8th, 1994, the Commissioners met for the first time in Delhi to design strategies and discuss professional plans, in cooperation with local victim and activist organizations. There were thirteen Commissioners from: United Kingdom (2); United States (3); China (1); Republic of Belarus (1); Poland (1); Germany (1); Canada (1); Italy (1); Sweden (1); and The Netherlands (1). Two of the Commissioners in the team were originally from India one from Bhopal itself. A young American physician who was working in India joined the group and provided additional help. The team worked in Bhopal for the following two weeks, and then returned to Delhi to evaluate their experience and plan for professional follow-up. The Commissioners divided their work in Bhopal into the following groups and areas of investigation:

1. Dispersion modelling: One commissioner collected information relating to the release and the atmospheric dispersion of the materials released from the plant to attempt a realistic model of the extent of exposure of the city to toxic materials.

2. Epidemiology Group: A questionnaire was administered to 479 persons randomly selected from the following categories:

- gas victims identified by the Indian government;
- slum dwellers near the Union Carbide factory;
- persons living five to eight kilometres from the factory;
- slum dwellers not exposed to the Carbide gases.

Within each group a stratified sample in age groups 18–35, 36–45 and 45–60 years was interviewed. Every fourth interviewee was referred to further investigation to the outpatient facility. Of the 124 selected for clinical examination, 96 attended and were medically examined.

3. Clinical Group: An outpatient facility was set up as a Clinical Research Unit in Kohe-Fiza which undertook the following clinical tests: respiratory system/lung function, neurological/neuropsychological function, immune system and ophthalmological status.

4. Evaluation of Family Life Quality: One Commissioner evaluated the impact and long-term consequences of the disaster on women and children (17 families), on their reproductive health, standard of living, household economics and community life.

5. Medical Care: Several commissioners met with local doctors, hospital administrators, the Red Cross, the Eye and TB hospitals, and the Gandhi Medical School Dean and faculty. A selection of medical records was examined and evaluated. The availability and quality of medical care was discussed with the victims.

6. Drug Therapy: The rationale of drug treatment and prescriptions was evaluated from the medical records of the gas victims and governmental therapy recommendations were studied.

7. Compensation Claims: Two Commissioners examined the laws and regulations relating to compensation claims, the claims court and the local procedures for identifying gas victims.

Findings of IMCB:

IMCB did not evaluate the factual and legal aspects of the Bhopal case. There is no doubt that the incident resulted from poor construction, maintenance and safety measures in running the Union Carbide factory. There had been previous incidents at the plant which should have made the management aware of problems within the company. No risk assessment procedure was undertaken for operating a factory with a 40-tons tank of highly toxic methyl isocyanate (MIC) in the close vicinity of a densely populated ward. The necessary information on hazardous substances used in the plant – in particular the nature of MIC – was lacking. Hazardous substances safety data sheets, mandatory in first world countries, were not made available to the local medical community. This added substantially to the absence of emergency plans for warning and evacuation of the population and the absence of a strategy for care in the acute phase.

The Tata Institute of Social Sciences (TISS) had conducted a survey on a large number of households in the affected wards, from the third day after the gas leak. This was commissioned by the State Government who are in receipt of the findings. TISS data are openly available. The Indian Council of Medical Research (ICMR) conducted a separate survey, from the third month, in areas severely, moderately and mildly exposed. Data collected by ICMR are not available and the results have not been published.

Clinical Investigation by IMCB

The objectives of the clinical study were:

- to describe the health status of Carbide gas-exposed people living in the affected areas;
- to determinate the prevalence of long-term health effects;
- to estimate exposure and enable modeling of dose-response relationships if possible.

Much of the early fatalities following the gas exposure was attributed to pulmonary damage. Because of the unknown composition of the "Carbide-gases", causalities might have been caused by other compounds, namely cyanide. It is known from eyewitness information by victims during the night of the disaster that different areas can be related to this assumption.

Post-mortem examinations of the fatalities during the first days showed extensive necrosis; examinations on those who died during the subsequent months showed the characteristic features of a small airways disease, *bronchiolitis obliterans.* The pathophysiology leads to long-term disabling respiratory disease, with shortness of breath, chest pain, chronic cough and eventually, given a latency period of 10 to 20 years, bronchial carcinoma.

Spirometry tests showed that lower mean values of the forced expiratory ratio were lower in the exposed groups. There was a trend of decreasing lung function with exposure. Of the 22 chest radiograms, eleven were considered abnormal. No lung cancer was detected.

The study shows the manifestation of typical lung dysfunction as a long-term damage due to MIC-gas exposure. The frequency and severity of this damage (mostly small airways) among individuals should be assessed. Since such lung disease is currently incurable, a systematic study to evaluate treatment including physiotherapy should be undertaken for management cases after spirometry tests.

Early treatment of inflammation and of superimposed infections with simple antibiotics is recommended. Equally important is the introduction of non-drug treatment strategies for physiotherapy and respiratory rehabilitation. A program for monitoring lung cancer should be instituted. The health-care system must be restructured to be prepared to deal with a large number of chronically disabled patients. The facts of the incurability and lasting nature of the pulmonary problems should be communicated effectively and widely. A programme on health promotion including the problem of tobacco smoking and environmental air pollution should be implemented. The hygienic condition of the wards requires urgent action. With the underlying lung problems the population is especially prone to other communicable lung disease.

Immune function testing:

Tuberculosis, a disease common to those economically disadvantaged and nutritionally and hygienically deprived, is a major health problem in India. Abnormalities of the immune system may be expected reflecting changes in the prevalence of this disease in the gas-affected population. Sixty-three subjects of the clinical cohort were tested by Mantoux tests, with 43 (75 percent) positive results. Fifty percent of the tested victims developed severe (3+ to 4+) reaction reflecting a high level of mycobacterium antigen sensitization.

Neurological and Neuropsychological Testing:

Very limited information is available on effects of the gas exposure on the peripheral and central nervous system. Frequent and consistent complaints of the gas-affected victims such as aphasia, loss of memory and concentration, exhaustion and weakness, chronic fatigue together with (unpublished) evidence on cerebral oedema and hemorrhage from post-mortem examination of deceased victims indicate the possibility of target-organ toxicity leading to central neuropathy in the surviving gas victims. A tier of neurological, neuromotorfunction and neuropsychological tests was applied. Loss of memory and disability to perform simple tasks was seen in a large proportion of victims. Symptoms and complaints have been previously denied or largely have been informally attributed to post-traumatic stress disorder. The government of India does not recognize post-traumatic stress as a compensable effect of the gas exposure.

Many patients had difficulties with finer motors skills and performed poorly in the line pursuit test; the striking results of this testing is the poor performance in the Benton short-term memory test. It is speculated that the exposures to the Bhopal gases are responsible for the observed central nervous deficiencies although some socio-psychological factors have been considered. Central nervous neurotoxic effects from the "Carbide-gases" as well as extended hypoxia might be responsible for this observation. Central neurological disturbance have been observed as resulting in organophosphate poisoning. The difficult social and housing situation of the cohort was evident during our visit to the wards in the affected areas. A possible confounder could be related to lack of manual and intellectual training, because many patients lost their work and could have reduced motor skills and memory.

Socioeconomic consequences of this illness adds emotional and physiological stressors to the victim's situation. Loss of livestock, loss of income-generating family members, loss of working ability or capacity results in a significant economic loss and burden to the people who were often struggling to survive prior to the accident. Young survivors complained of decreased performance at school sometimes leading to many dropping out due to their symptoms.

Pharmacotherapy:

The lack of studies on therapies given for the injuries sustained by the gas victims, raised concern about the rationale and efficacy of current treatments. In 1990, a study by one of the victims' organizations on 522 patients at gas-relief hospitals found that prescriptions were irrational, including unnecessary prescriptions and hazardous drugs which were banned in other countries. Seventy-two percent of the patients found that their medicines were ineffective.

An IMCB survey of drugs kept in the households, prescribed by the slum-area hospitals or local doctors and kept in the pharmacies, revealed that most treatments were given for short periods for symptomatic relief of pain, shortness of breath and respiratory infection. There was little specificity as to choice of therapies, several compounds had the potential for toxicity and iatrogenic disease, especially with corticosteroids prescribed in high doses without appropriate timed regimens.

The development of therapeutic guidelines and health education programmes as a specific intervention is recommended. A community primary health-care system could be an appropriate instrument for both the task of education and monitored therapeutic trial.

Health Book:

During the epidemiological and clinical study we had the opportunity to look at records of previous medical examinations and findings the patient had kept. The poor quality of the documentation of the victims' past medical history prompted IMCB to prepare for all persons studied individual health record books, containing the results of the health questionnaire and of clinical, biochemical and radiographic investigation and prescriptions, as well as treatment proposals. It is anticipated that a record of the patients' history kept by the patients will enforce adequate documentation of medical findings and treatments including rational, diagnosis-based pharmacotherapy adding to the improvement of health care of the patients.

Compensation:

Following a disaster, health authorities must first focus on provision of care for the affected and injured. They must, however, also be prepared to conduct surveillance and epidemiology. Understandably, the primary care for the victim of the accident had the highest priority. However this short time period is also best for collection of adequate information that becomes invaluable in determining the acute and chronic health effects, and full- and long-term impact of the disaster. These data are needed for:

- identification of exposed and ill patients;
- provision of adequate short- and long-term care and surveillance;
- determination of short and long term health consequences;
- prevention of future disasters;
- improvement of contingency planning and mitigation of future disasters;
- linking exposure and effects for litigation and compensation.

In Bhopal nobody was prepared for the disaster and it is obvious that because of practical impediments nobody was prepared to collect information that otherwise could be made readily available. The night of the disaster, an enormous flood of patients came to the hospitals. Even the most basic medical documentation was not obtained for most of the patients. Practical constraints ranged from unavailability of death certificates and medical records, to absence of prior census data, environmental data and equipment, and limited epidemiological expertise to collect such data. Determining health effects attributed to a disaster requires exact estimates of exposure; in Bhopal this was not obtained on MIC or its breakdown products. The use of "distance" or "damage to vegetation" as surrogates for exposure assessment offers only simplified and imprecise approximations. Studies conducted during months following the disaster on lung function and ocular problems were not followed up or, if done, not published. Thus there has not been much information added to what was known from the first weeks of the disaster. This lack of data holds its own tragedy by denying to affected individuals, especially children, proper restitution, rehabilitation and compensation.

Recommendations of IMCB:

IMCB hopes that the Bhopal tragedy will become an important tool for promoting the medical and environmental awareness of industrial and environmental hazards, and recognition of the challenges they represent to the role of the medical system.

1. The Commission has found genuine long-term morbidity in a substantial proportion of the population. The Commission recommends that the spectrum of disease attributed to the release and subsequent exposure to the "Carbide-gases" be broadened to include neurotoxic and post-traumatic stress disorder.

2. Bhopal's current, hospital-based delivery of health-care is inappropriate for the chronic nature of the disease. Victims with permanent health problems require continuing care and rehabilitation. It is recommended that the Indian health authorities adopt Bhopal as a priority for the implementation of their stated policy of community health-based approaches.

The current needs of the affected population are different from those in the earlier phases of the tragedy. Priority should be given to a network of community-based clinics which would more equitably and efficiently provide routine care for the population while avoiding unnecessary pressure at the hospital level. Hospital resources should be reoriented, mainly on a referral outpatient basis. Carefully planned and documented interventions must be developed and implemented, mainly rehabilitation, and especially physiotherapy and pharmacological strategies. The Commission found evidence of irrational, unnecessary and costly drug treatment, which leads to further exploitation of the gas victims.

The Commission learned that many families spent a large proportion of their income on drugs prescribed by local doctors. There is no specific treatment and the promises of western wonder drugs are unjustified. It is necessary to explain to the affected that the long-term lung damage is incurable. It must be stated that the medical system which lets private doctors practise in Bhopal without quality control must be blamed for inappropriate polypragmasia. Analgesics, antibiotics, antihistamines and bronchodilators are widely prescribed without the patient being properly informed about his diagnosis and the drug regimen. Chronic conjunctivitis related to dryness of the eyes was prevalent; local physicians are using expensive antibiotic and corticosteroid formulations instead of appropriate installation of a simple and cheap saline solution.

3. The Commission has found that most of the data collected by the Indian Council on Medical Research (IMCR) and others is not freely available. This has resulted from a dissociation between research studies on the population and health care delivered to the population. The Commission recommends an urgent review and dissemination of such data. Further information should be collected in the context of routine health-care delivery and should include an evaluation of the current and long-term effects on women and children.

4. The Commission recommends that an independent group be established, including victims' organizations, the Indian government, and Indian and international experts, to examine the criteria for assessing

disability; in the light of emerging international consensus this should include socioeconomic as well as strictly medical consequences.

5. The Commission recognises that Bhopal is a tragic model of an industrially-induced epidemic. The Commission has decided, based on its experience, to organise international teams, when requested, to provide technical assistance and on-site evaluation of industrial/environmental disasters for the benefit of affected communities.

Source: Results of the International Medical Commission on Bhopal (IMCB)
Birger Heinzow, MD

Conclusions of Indian Council of Medical Research epidemiology study (1996)

A study by the Indian Council of Medical Research (ICMR) that reflects the challenge of determining health impacts in Bhopal through epidemiological methods, but that nonetheless determines that exposed people are still dying faster than the unexposed.

The following are some of the noteworthy features that have emerged from this epidemiological study.

1) At the outset, it may be pointed out that the age and sex distribution of the population of "affected" as well as "control" areas were almost similar comparable to national population pyramid.

2) A noteworthy feature was that the "death rates" were higher in the "exposed areas", than in the "control areas", throughout the ten years period of observations.

3) The "gas exposure" particularly in the severely affected area showed higher mortality in the initial years, which gradually declined and nearly touched "local" or "national levels." Deaths in the exposed area were mainly due to respiratory disorders throughout the period of observations. Death rates were higher in the age group of 45 years and above.

4) Another notable feature was the "pregnancy rate," which is generally associated with disasters in general. The rate was high till 1986–87 and gradually declined over a period of time. Likewise, by 1989 the "abortion rate" in the affected areas, which was initially 12%, declined to about 7.5%, as against 1.4% in the control area. Such phenomenon has been observed in man-made and even natural disaster.

5) General morbidity as well as that traceable to respiratory or ophthalmic morbidity, based on the symptomatology reported by the patients or the responsible family members, was observed to be consistently higher

in affected areas as compared with the control areas. The "immediate" morbidity was about 95–97% for both pulmonary and ophthalmic involvement. But there was no rapid drop within a short term of 2–3 months. However, the eye condition worsened once again later on. Interestingly enough, in the last phase beyond 1992, ocular morbidity was higher in the mildly affected areas.

6) Thus, from the analysis of the data on the effect of toxic gas exposure on health and review of the meager literature, especially the nature of the Bhopal gas exposure, it is obvious that apart from the immediate raised mortality, there was persistence of morbidity in the affected areas existing over 10 years of study.

7. Based on the epidemiological study the following recommendations have emerged:

i) In any widespread chemical disaster, relief measures should be taken for the victims without delay. Simultaneously, steps should be taken from the earliest to launch well-planned epidemiological studies to monitor mortality rates and morbidity status of affected population.

ii) Detailed steps should be taken to ensure that the natural history and the evolution of disease entities is finally characterized.

iii) The consequence of extensive chronic pulmonary disease in the wake of chemical accidents should be investigated for functional rehabilitation and restoration of physical function and affected systems. Wherever necessary, norms for the clinical management and relief of airways obstruction with the state of the art instrumentation and physiotherapy should be accepted simultaneously.

iv) In the unfortunate event of lack of information on the exact nature of the chemicals or their toxic metabolites, and the specific antidotes, symptomatic relief drugs only remain the mainstay of clinical management.

v) It would be desirable to extend the long-term study of the same cohort in Bhopal to study in the potential hazards of cancer and long-term involvement of other organs.

Conclusion

1. The Bhopal Gas Disaster was entirely man made because: (a) If the Union Carbide pesticide factory did not have any habitation within 4 km radius, nobody would have been hurt; (b) If methyl isocyanate had not been stored for such a long time (rather unusual) the accident could have been averted; (c) If the monitoring and safety devices had been maintained, the accident would have been averted.

2. The toxic gas to which the Bhopal population was exposed consisted of MIC and a vast amount of its reaction products inside Tank 610.

3. The mortality and morbidity caused by the toxic gas(es) inhalation was a "one time acute injury" to the respiratory tract and the ophthalmic system and which often healed with resolution or necrosis and fibrosis, but did not lead to progressive pulmonary or ophthalmic disease resulting in blindness. The scars produced after the acute lung injury and their sequelae may however, continue to produce recurrent episodic respiratory illness and possible disability because of secondary respiratory infection and airway hyper reactivity or fibrosis, emphysema, bronchiectasis etc. for a long time or even the whole life. People with preexisting lung disease (presumed at least 5% in any population), or smokers, after the gas exposure would have suffered more than those who were healthy before the exposure.

Source: Bhandari N. R., et al. "Reproductive Outcomes Subsequent to Gas Leak in Bhopal: An ICMR study. *Indian Journal of Preventive & Social Medicine. 1996 July–Dec; 27(3&4): 45–51.*

Mental health impact of Bhopal gas disaster

A report on the mental health situation among the victims, by an expert involved with treatment and research in the aftermath of the disaster.

Summary

During the night of December 2-3, 1984 the world's worst industrial disaster took place in the city of Bhopal in central India. Large amounts of toxic gas leaked from the plant into the surrounding area, which was densely populated. More than 2,000 died immediately and over 200,000 population were directly affected in a city of 700,000 population. The disaster-affected populations have been investigated for the effect of the disaster on their physical and psychological health. Community level studies carried out within one month of the disaster to 10 years after the disaster report higher levels of physical and mental health morbidity. Though efforts to provide psychological support to the affected population were initiated using the primary care personnel by focussed training programmes, a system of comprehensive community based health care in general and mental health care in particular, is still not in place. In addition there is need for continuing the research studies into the long-term effects of the disaster and the morbidity in the affected population. The magnitude of the Bhopal disaster and the research efforts to understand the health effects have resulted in greater awareness in India of the psychological aspects of disasters and to include psychological support as part of relief and rehabilitation activities following all disasters.

Impact on Mental Health

Bhopal disaster is the first disaster in India to be studied systematically for the mental health effects. Information is available about the mental health effects from a number of sources. These are from studies as part of general health surveys as well as specific studies on mental health. The direct involvement of the psychiatrists/neurologists at the field level did not occur till about 8 weeks after the disaster. This delay was in spite of the recognition of the importance of mental health effects of the disaster within the first fortnight of the disaster. By coincidence the Fourth Advisory Committee on Mental Health of ICMR was meeting on December 12-14,1984. The experts in the meeting recognized the need of the affected population as follows:

"The recent developments at Bhopal involving the exposure of 'normal' human beings to substances toxic to all the exposed, and fatal to many, raises a number of mental health needs. The service needs and research can be viewed both in the short-term and long-term perspectives. The acute needs are the understanding and provision of care for confusional states, reactive psychoses, anxiety-depression reactions and grief reactions. Long term needs arise from the following areas, namely, (i) psychological reactions to the acute and chronic disabilities, (ii) psychological problems of the exposed subjects, currently not affected, to the uncertainties of the future, (ii) effects of broken social units on children and adults, and (iv) psychological problems related to rehabilitation."

Mental Health Studies:

The initial assessment in the first week of February 1985 (about eight weeks after the disaster) was done by R. Srinivasa Murthy (RSM), of the National Institute of Mental Health and Neuro Sciences (NIMHANS), Bangalore, and Professor B. B. Sethi (BBS), of K.G. Medical College (KGMC), Lucknow. They visited the city and interacted with the general population, the patients attending the health facilities and the medical personnel, to understand the magnitude and nature of the mental health problems in the affected population. Their observations, over a week's time, were based on clinical and unstructured interviews. These initial observations led to an estimate of the magnitude of mental health needs of the population at 50% of those in the community and of about 20% of those attending medical facilities (Srinivasa Murthy, 1990).

Immediately following these observations, during February-April 1985, a KGMC team carried out systematic studies. As a first step, ten general medical clinics in the disaster-affected area were chosen. A team consisting of a psychiatrist, a clinical psychologist, and a social worker visited one clinic a day, by rotation in a randomized fashion, on three occasions and screened all the newly registered adult patients with the help of a self-reporting questionnaire (SRQ). Subjects identified as probable psychiatric patients were then evaluated in detail by the psychiatrist with the help of a standardized psychiatric interview, the Present State Examination (PSE). Clinical diagnoses were based on the International Classification of Diseases (9th revision) (ICD-9) (WHO, 1975).

During a period of 3 months (February-May 1985), of the 855 patients screened at the 10 clinics, on the basis of their SRQ scores, 259 were identified as having a potential mental disorder. Of these potentially mentally ill people, 44 could not be evaluated, and 215 were assessed using the PSE. The confirmed number of psychiatric patients was 193, yielding a prevalence rate of 22.6%. Most of the patients were females (8.11%) under 45 years of age (74%). The main diagnostic categories were anxiety neurosis (25%), depressive neurosis (37%), adjustment reaction with prolonged depression (20%), and adjustment reaction with predominant disturbance of emotions (16%). Cases of psychosis were rare, and they were not related to the disaster.[9]

During the same period, in the third month of the post-disaster period, neurological studies were carried out. This was a survey of the gas-affected patients admitted to the various hospitals in the Bhopal city. A total of 129 adults and 47 children were studied for neurological problems. Evidence of involvement of the central nervous system was present in three patients in the form of stroke, encephalopathy and cerebellar ataxia. Affection of the peripheral nervous system was observed in 6 patients. Vertigo and hearing loss occurred in 4 patients. Many patients reported transitory symptoms like loss of consciousness (50%), muscle weakness, tremors, vertigo, ataxia and easy fatigability. Most of these symptoms cleared up after varying periods of time. Of the 47 gas affected children, loss of consciousness at some time or other occurred in half of the patients. Fits occurred during the course of the illness in 3 children. Mental regression was observed in one child who had commenced speaking in sentences but stopped talking after the disaster. There were no abnormalities in the neurological examination in all of the children. An important observation by the doctors who had examined the children during the early phase of illness was generalized hypotonia and weakness. Two children were noted to be "floppy" with weakness of limb movements and had difficulty in getting up from the ground. Of the 3 patients who had central nervous system involvement, the patient with stroke died. His autopsy showed intense congestion and petechial hemorrhages of the gray and white matter with frank hemorrhage in the circle of Willis area, perhaps indicating the sustained microvascular damage by the circulating MIC.

Subsequently, from June 1985, the Lucknow team with the funding from ICMR, New Delhi team conducted a detailed community-level epidemiological study, along with the community level epidemiological study for other health effects. This study included recording of the complaints of subjects, and the record of illnesses and deaths in 100,000 population in the different areas of Bhopal. A fresh census of the total population was undertaken prior to the study. The sampling frame was drawn in such a manner that populations variously exposed to the disaster were included along with a control group located far away from the gas-exposed area, but from the city itself.

The methodology used for screening of the households was interview with the head of the household for the presence of symptoms from a standardized checklist. Those found to have symptoms were further seen by a qualified psychiatrist who administered a detailed mental status examination instrument (PSE-9th version) and arrived at the ICD-9th version diagnosis. Each year a new set of families were sampled and studied in addition to follow-up of the patients diagnosed in the previous years.

The results of the first-year survey involved 4,098 adults from 1,201 households. A total of 387 patients were diagnosed to be suffering from mental disorders, giving a prevalence rate of 94/1,000 population. Most of the population consisted of females (71%); 83% were in the age group 16-45 years. Ninety-four percent of the patients received a diagnosis of neurosis (neurotic depression, 51%; anxiety state, 41%; and hysteria, 2%) and had a temporal correlation with the disaster. For the next three years, the team repeated the annual surveys and follow-up of the initial patients identified by the community survey. Detailed case vignettes and descriptive accounts of the patients from the Bhopal disaster were prepared.

These general population psychiatric epidemiological studies show that the gas exposed population were having significantly higher prevalence rates for psychiatric disorders in comparison to the general population. The gradient relationship of higher rates of psychiatric morbidity with severity of exposure to the poisonous gas was maintained throughout the 5 years of the survey period. At the end of the five-year period the number recovering fully was small and large numbers continued to experience the symptoms along with significant disability in functioning.

End notes and references:

9 Wing, Cooper & Sartorius, 1975

Source: R. Srinivasa Murthy, "Mental Health Impact of the Bhopal Gas Disaster," Fact Finding Mission on Bhopal.

Methyl isocyanate exposure and growth patterns of adolescents in Bhopal (2003)

Conclusion of research letter (coauthored by Nishant Ranjan, Satinath Sarangi, V.T. Padmanabhan, Steve Holleran, Rajasekhar Ramakrishnan, and Daya Varma), examining the effects of exposure to Carbide's gases on the physical growth pattern of offspring of exposed people.

We found selective growth retardation in boys, but not in girls, who were either exposed as toddlers to gases from the Bhopal pesticide plant or born to exposed parents. The fact that exposed and unexposed girls were virtually identical in all measures suggests that the exposed and unexposed groups were well-matched and that the association observed in boys is truly a result of exposure and not of other unobserved differences in the demographics. The main chemical that escaped from the plant was MIC, which is readily degraded on contract with water and in the body. One of the degradation products of MIC, is trimethylamine, which has been reported to produce selective growth retardation of male progeny of mice, associated with a decrease in serum testosterone. It is possible that similar hormonal effects were produced by MIC, its metabolites, or other substances.

Source: Summarized from the *Journal of the American Medical Association*, October 8, 2003, p. 1857.

8

Contamination

Water contamination, the next crisis

SINCE 1990 IT has been clear that effluent from the former factory is contaminating the soil and ground water nearby. This story of water contamination, although less dramatic, has followed the same unfortunate pattern as the gas leak. As shown in this chapter, the chemical disposal system in the UCIL Bhopal plant had a lower safety rating than its American sister plant from the design stage. The company, according to its own internal documents, had known that the water was heavily contaminated since 1989. Its policy today veers between denial of the fact of contamination and denial of the responsibility to remediate the site.

Union Carbide's position today is that "the [Indian] government tested it [the water] and found nothing wrong with it" (Letter from Tomm Sprick, 18 November 2004). The company does not assert that there is no contamination, only that certain reports – for example, the 1997 government report of the National Environmental Engineering Research Institute (NEERI) – say that although there was soil contamination, it would take twenty-three years to reach ground water. However, there are many other studies, including Carbide's own, that have concluded otherwise.

Given that ground pollution clearly does exist, and that the carcinogenic, industry specific chemicals and heavy metals that have been found in the water clearly originate from the factory site (as there is no other comparable industry in the vicinity), Carbide, and now Dow's, fall back position is that "while we have seen conflicting reports currently being made by various groups and media, we have no firsthand knowledge of what chemicals, if any, may remain at the site and what impact, if any, they may be having on area groundwater. We believe it is important for the State of Madhya Pradesh to restart and complete the remediation of the plant site. The state is in best position to evaluate all available scientific information and to make the right decision for Bhopal. For specific details, you'll need to contact the government of Madhya Pradesh" (Sprick to Hanna, 2004). Custody of the site was transferred to the Madhya Pradesh government in 1998; in the fourteen years that Carbide held the site after the disaster, they did not complete the remediation, and now expect Indian taxpayers to take up the responsibility. The condition of the site has been documented by Greenpeace in a report excerpted below.

There has, again, been very little research into the effects of the water contamination on health – this chapter begins with one such limited study conducted by the Sambhavna clinic. Despite the 7 May 2004 directions by the Supreme Court of India to the state of Madhya Pradesh to begin the supply of clean water to the residents of the affected communities, currently, according to community residents, no more than ten percent of necessary water is being delivered. Although the contamination issue was brought by activist groups to the attention of both Union Carbide and the central Indian government as early as 1990, at this writing in 2005 the communities that absorbed the brunt of the gas disaster, are still drinking dangerous water that contains very high levels of heavy metals and inorganic chemicals.

There is currently a lawsuit on behalf of the survivors and organizations in the Federal District Court in New York, before Keenan – the same Judge who refused to allow the civil suit against Union Carbide to be tried in the U.S. – that seeks to compel Union Carbide to remediate the Bhopal site, and to compensate those whose health has been adversely affected by the migration of that water into the underground aquifers. This litigation, brought by lawyer H. Rajan Sharma, is described in the chapter "Just Compensation?".

This chapter also includes Kalpana Sharma's analysis of the reasons why it took an extended hunger strike by Bhopal survivors and activists to extract a statement from the Indian government before the U.S. District Court that it would have no objection to the Court ruling that the Union Carbide Corporation be compelled to clean up the contaminated factory site. Her article also shows the intransigence and insensitivity that the Indian government has brought to the Bhopal issue since the disaster in 1984.

Union Carbide's internal documents regarding groundwater contamination (1982–1997)

Carbide's internal documents clearly show that the company knew there was a possibility of contamination of the groundwater from the beginning, knows there is currently contamination at the site, and has known all this for some time.

1982

Differences between Bhopal plant and Institute plant:

Environmental Impact Assessment states that the Bhopal plant is modeled on Union Carbide's plant at Institute, West Virginia but only "[w]here suitable" (UCC 04204).

The Institute plant had high EIA ratings "based on water discharges into the Kanahwa River" whereas "process design for Bhopal is based on no discharges to surface waters. All wastewater streams from the Pesticide Unit at Bhopal will discharge into solar evaporation ponds. All wastewater will be contained in these in-plant ponds."

UCC Engineering Department warns of danger of groundwater contamination even at the design stage (July 21, 1972):

Proposed design poses "danger of polluting subsurface water supplies in the Bhopal area" and "new ponds will have to be constructed at one to two-year intervals throughout the life of the project" in order to address this problem (UCC 04129).

UCC approves project with all these design problems and environmental risks:

Project approved for $20 million (UCC 04240).

"Impermeable" linings of solar evaporation ponds leaked in Bhopal during plant operations:

Telex dated March 25, 1982: "Phase II evaporation pond almost emptied. Reps of K. R. Datey at site and investigation of the leakage in progress. Unfortunately, emergency pond has also shown some signs of leakage." (UCC 01737)

Telex dated April 10, 1982: "Continued leakage from evaporation pond causing great concern." (UCC 01736)

1990

UCC takes a decision to appoint Arthur D. Little, Inc. as primarily responsible for all aspects of site rehabilitation efforts:

"In view of the above, it would be necessary to seek assistance and advice of an expert organization having firsthand experience in this field. Since no Indian organization has had similar exposure, it has been decided to appoint M/s A. D. Little & Co. of USA which has considerable experience in this field.... At the instance of the M.P. State Govt, it is proposed to appoint National Environmental Engineering Research Institute (NEERI) for carrying out above investigation under the overall guidance of M/s A. D. Little & Co." (UCC 02271)

UCC lies to M. P. Government about NEERI's 1990 study:

NEERI's 1990 results according to UCC show "no contamination of soil and ground water" existed at UCIL site. (UCC 03485)

UCC "Business Confidential" docs show that UCC knows NEERI's 1990 data does not show that:

"While the ponds were clearly the focus of this [NEERI] study, the close proximity of the ponds to the plant, relative to the 10 km radius, seems to implicitly 'clear' the plant site itself.... However, I would advise caution in using the NEERI data, for two reasons: 1) the study was done for the state government, and I am not sure whether they are ready to publish it broadly, and 2) we do not know the exact sample and analytical protocols used by either group." (UCC 02050)

UCC's secret, internal studies show massive soil and water contamination in plant site:

"Presence of Toxic Ingredients In Soil/Water Samples Inside Plant Premises."

"The seriousness of the issue needs no elaboration. It is earnestly suggested that the subject be given due consideration and studies initiated without further delay."

"Samples drawn in June–July '89 from landfill areas and effluent treatment pits inside the plant were sent to R and D. They consisted of nine soil/solid samples and eight liquid samples. The solid samples had organic contamination varying from 10% to 100% and contained known ingredients like naphthol and naphthalene in substantial quantities.

"Majority of the liquid samples contained naphthol and/or Sevin in quantities far more than permitted by ISI for on-land disposal. All samples caused 100% mortality to fish in toxicity assessment studies and were to be diluted several fold to render them suitable for survival of fish." (UCC 02268)

UCC is still arguing in Court that, based on NEERI's 1997 report, there is no contamination in or around the Bhopal plant:

UCC states in its motion papers that "there was no groundwater contamination outside the plant" due to the "relative impermeability of the soil in and around the plant." (Def. St. at ¶ 6; Ex. A to the Krohley Declaration.)

Even though UCC knows full well that its own consultant, Arthur D. Little (ADL), specifically rejected the conclusion in NEERI's 1997 report:

ADL's comments on NEERI's 1997 Report were:

"2. Groundwater Issues: There are two major issues we have identified concerning groundwater at the site:

"2.1 Statements concerning contaminant travel times to the aquifer below the site should be considered highly speculative. There is very little site-specific data that can be used to confidently predict infiltration rates. The information that does exist suggests that travel times could be significantly less than identified in the report. Refer to Tier II Comment No. 41 for details.

"2.2 There does not appear to be sufficient information to discount a potential impact to groundwater from contaminated soils present on the facility…

"If remedial action is completed as quickly as possible, the potential for contaminant migration from soil to ground water will be diminished significantly." (UCC 03032)

"The conclusions regarding travel time to the water table may significantly underestimate the potential for groundwater contamination... However, site-specific data from the report suggest that travel times could be significantly faster than assumed." (UCC 03042)

"As an example, one can argue that the worst case scenario travel time would be 2 years…" i.e. from 1997 (UCC 03043).

After the lease was surrendered to State, M. P. authorities wrote to UCIL demanding that Union Carbide clean up its mess:

"As per rules M/s.Union Carbide are fully responsible for the environmental remediation of the problem created by them. It is also the responsibility of the administration to get the above land decontaminated. M.P. Pollution Control Board's responsibility is limited to monitoring and to see that environmental rules are followed.

Therefore it is prayed for the Hon'ble Court's direction to Industries Department to get the work of remediation of all of the above environmental contamination done by Union Carbide Bhopal because under the Hazardous Waste (Management & Handling) Rules 1989, 594 (E) Section 3 Subsection (1) and Section 4(1) whoever has produced the contaminated waste it is his responsibility to decontaminate it. Therefore as per rules it is the responsibility of Union Carbide Bhopal to pay for all the expenses being incurred on all the above work." (UCC 02237)

But UCIL (now Eveready), and UCC had told M. P. State that it was not possible:

"The company ceased to be the occupier of the site on and from 9.7.98. The State Government as the rightful occupier of the premises and having full knowledge of the status of the site is expected to do whatever is required to be done in regard to the site... The company is neither in a position nor is required to be further involved in the various activities which the State Government as the occupier may think it fit to undertake now by itself or through any of its agencies." (UCC 02240).

Source: Union Carbide Corporation discovery documents, excerpted from www.bhopal.net.

M.P. Public Health Engineering Department's report on the presence of chemicals in the groundwater in the vicinity of the Union Carbide factory (1996)

In 1991 and 1996 tests on local groundwater taken from 11 tubewells were carried out by the M.P. Public Health Engineering Department's State Research Laboratory. Extract of report translated from the original in Hindi.

"On 26.11.'96, ten samples were collected from J. P. Nagar, Atal Ayub Nagar, Arif Nagar, Chhola and Kainchi Chhola, all situated close to the Union Carbide factory...

"All samples were subjected to both bacteriological and chemical analysis. The results show that the groundwater is contaminated with bacteria and there is a heavy presence of chemicals. Normally the COD (Chemical Oxygen Demand) value in ground water is zero but the samples tested here had COD values between 45 mg/l and 98 mg/l whereas, the WHO has fixed the standard value of COD for natural water at 6 mg/l. The high values of COD found in the groundwater establish that large amounts of chemicals are dissolved in it.

"Usually COD cannot be brought down by commonly used techniques. When river water is contaminated with chemicals one has to wait for it to come down and this problem is controlled in a few days by dilution. With ground water such a solution is not possible hence it will be proper to stop these sources.

"Water from tubewells in other parts of Bhopal were examined at this laboratory. However, chemical contamination was found only in these areas. The tubewells in these areas were tested five years back and at that time too the results showed chemical contamination. Hence, it is established that this pollution is due to chemicals used in the Union Carbide factory that have proven to be extremely harmful for health. Therefore the use of this water for drinking must be stopped immediately."

Sd. Chief Chemist
State Research Laboratory
Shyamla Hills
Bhopal

The National Environmental Engineering Research Institute (NEERI) report (1997)

Indian government research agency NEERI's report titled "Assessment of contaminated areas due to past waste disposal practices at EIIL, Bhopal," was sponsored by Eveready Industries India Limited (corporate successor to UCIL) and is often cited by both Dow and Carbide. Following is an extract from the report.

Groundwater

Seventeen ground water samples collected in and around EIIL do not show the presence of semi volatiles, organics, heavy metals and inorganics. The water meets the drinking water quality criteria. This indicates that the contaminants have not reached the water table till now. In general, the soil in the area is clayey soil with more than 45% clay content. This clayey soil is highly impermeable and would travel approximately at the rate of 36 cm/year at the permeability rate of 1×10^{-5} cm/sec i.e., it would take 23 years for the contaminants to reach the groundwater table provided the leachate does not find a channel to migrate at a faster rate. This could be the reason for the water not getting contaminated.

Source: *Assessment of contaminated areas due to past waste disposal practices at EIIL, Bhopal*, By the National Environmental Engineering Research Institute, Nehru Marg, Nagpur 440 020, October 1997
Sponsored by Eveready Industries India Limited (EIIL), Bhopal.

Arthur D. Little's (ADL) comments to Carbide on the NEERI report (1997)

ADL, the consulting firm that Union Carbide depended on to craft their "sabotage theory" report, had these criticisms of the 1997 NEERI report that are UCC's primary defense against allegations of continuing groundwater contamination.

2. Ground Water Issues: There are two major issues we have identified concerning ground water at the site:

2.1 Statements concerning contaminant travel times to the aquifer below the site should be considered highly speculative. There is very little site-specific data that can be used to confidently predict infiltration rates. The information that does exist suggests that travel times could be significantly less than identified in the report. Refer to Tier II Comment No. 41 for details.

2.3 There does not appear to be sufficient information to discount a potential impact to groundwater from contaminated soils present on the facility...

Source: UCC Internal Documents, courtesy Sambhavna Documentation Center, Bhopal.

Madhya Pradesh Pollution Control Board (MPPCB) groundwater analysis (2004)

This summary of ground water samples collected around UCIL premises (April 03–Jan 04) found a variety of chemicals and pesticides consistently in wells within 2 km radius of the former factory.

The Board has collected groundwater samples around UCIL premises [within 2.0 km radius of factory premises] from the following locations in Apr 03: Atal Ayub Nagar, New Arif Nagar, Arif Nagar, Annu Nagar, Kainchi Chhola, near Dussera Maidan, Gareeb Nagar, Kainchi Chhola Gali No. 3, near Ujjain cabin, Blue Moon Colony, Preet Nagar, Solar Evaporation pond and Rajeev Nagar. The analysis of these samples reveals that the parameters viz. colour, turbidity and chlorides of some samples exceed the desirable limits of BIS*– 10500 whereas parameters viz. total hardness, total alkalinity, DS** and fluorides exceed the said limits in most of the samples. Pesticides like Lindane, Aldrin and B-BHC were detected in some of the samples.

In July, 03 the ground water samples were collected from the same locations. The analysis of these samples reveals that the parameters viz. colour, turbidity and chlorides of some samples exceeds the desirable limits of BIS –10500 whereas parameters viz. total hardness; total alkalinity and DS exceed the said limits in most of the samples. Pesticides like Lindane, Methoxychlor, Heptachlor, Aldrin and Dieldrin were detected in some of the samples.

In Oct, 03 the ground water samples were around UCIL premises and collected from the same locations. The analysis of these samples reveals that the parameters viz. colour and chlorides of some samples exceed the desirable limits of BIS – 10500 whereas parameters viz. total hardness, total alkalinity and DS exceed the said limits in most of the samples. Pesticides like Lindane, Heptachlor, Aldrin, Dieldrin, BHC, Endrin and 4,4 DDT were detected in some of the samples. Halogenated hydrocarbon viz. 1,2,3 TCB was detected in some of the groundwater samples.

In Jan, 04 the ground water samples were around UCIL premises and collected from almost the same locations. The analysis of these samples reveals that the parameters viz. chlorides and fluorides of some samples exceed the desirable limits of BIS –10500 whereas parameters viz. turbidity, total hardness, total alkalinity and DS exceed the said limit in most of the samples pesticides analysis in process.

* Bureau of Indian Standards

** Dissolved Solids

Source: From the report of monitoring of groundwater quality by the Madhya Pradesh Pollution Control Board, unpublished 2004.

Findings of survey of Annu Nagar (groundwater contaminated) (2003)

In one of the only efforts to document the health effects of ground water contamination from the Union Carbide factory, the Sambhavna clinic surveyed 1528 individuals from 270 families, between 17 July 2001 to 9 January 2003. This is the brief conclusion:

According to a recent survey carried out by the community health workers of the Sambhavna clinic in Annu Nagar (population 1528), 91%

of the residents were using water from the contaminated hand pumps. According to the survey, every second person in the community was suffering from a multitude of symptoms. The most common symptoms among all age groups were found to be abdominal pain followed by giddiness, pain in chest, headache and fever. These symptoms were most frequent among gas affected people who were additionally exposed to contaminated water. One of the significant findings of the survey was that among the teenage females between 13 and 15, 43% had not begun their periods.

Source: Sambhavna Clinic Documentation Center, 2003.

The Greenpeace study of chemical stockpiles at the Carbide plant (2002)

In 2002 Greenpeace International took a scientific team to the grounds of the former Union Carbide plant to record and measure the chemical stockpiles left behind at the site, and their continuing effects on the local environment.

1.6 Conclusions

The current survey represents the most comprehensive and detailed description available on the chemical stockpiles at the UCIL facility. The combination of quantitative and qualitative analyses allows a broad understanding of the contents of the individual stockpiles. Organic compounds detected in the solid wastes left unattended and insecure on the territory of the former UCIL plant are variously toxic, persistent and/or bioaccumulative. Confirmation of contamination outside the factory walls is moreover provided by the four samples from the SEP area.

In terms of immediate threat to human health the carbaryl and BHC must be regarded as of the greatest concern. These were found at almost every site sampled. The presence of sevin, widely regarded as insufficiently persistent to remain so many years after the abandonment of the plant, is notable. Whilst its biodegradability does mean that there is limited risk of dispersal to the wider environment, the fact that concentrations are low does not necessarily preclude the presence of more heavily contaminated materials and there remains the risk of exposure to any person or persons who may come into contact with the stockpiled wastes. The toxic action of carbaryl results from its inhibition of an enzyme critical to regulation of the passage of signals between nerve cells. Large doses can be fatal to humans unless an antidote is administered.

The hazard posed by the carbaryl is exacerbated by its combination with HCH isomers which were found in concentrations ranging from part per million to percent levels. These isomers have differing modes of toxicity. Gamma-HCH (lindane) is the member of the group for which toxicological actions are best understood. Human deaths have only rarely been recorded subsequent to lindane exposure but are not unknown. Its toxic actions result from nervous system stimulation (IPCS 1991, Smith 1991). However, alpha-HCH, which predominates in almost all samples, is a nervous system depressant, as are the beta-and delta isomers (Willett et al. 1998, Smith 1991). The combined effects of these and the other contaminants is consequently extremely difficult to predict.

In addition to the immediate threat to health, HCHs, once absorbed in the body, can be retained for years and have the potential to cause long-term health effects. HCHs are also reasonably anticipated to be human carcinogens by the U.S. Department of Health and Human Services (DHHS 2000); the International Agency for Research on Cancer regards them as possible human carcinogens (Group 2B). HCH isomers are persistent not only in the body but in environmental systems as well.

The contamination of the soil around the UCIL site will therefore be expected to represent a long-term issue that could be significantly exacerbated by the continued spread of materials from the stockpiles given the perilous state of their containment.

Furthermore, under Indian climatic conditions they can evaporate quite quickly (Samuel et al. 1988) and these volatilized residues may be transported to distant areas and contribute to global atmospheric contamination. This dispersal mechanism has over recent decades caused widespread contamination of food products and most human exposure can take place through consumption of contaminated food products or through transfer from mother to child transplacentally or via breast-feeding (Willett et al. 1998). Many of the other contaminants, notably the other organochlorines, are anticipated to behave in a similar fashion.

In the context of the UCIL site, routes of population exposure to these chemicals include inhalation of or dermal contact with contaminated dusts on the site or blown on the wind and consumption of dairy products from cattle grazing in and around the site, in addition to the more conventional routes. The easy access to the site and sheds and buildings within it also raises the possibility of material being removed, distributed and accidentally misused. Children playing in the site may also come into direct contact with any of the unsecured material described in this report.

Exposure of the local population to the contaminants buried in the former evaporation ponds must also be considered. The plastic liner has been breached in at least three locations and the samples collected as part of this survey show traces of chlorobenzenes in the soil at this location. Local residents are known to take soil from here to use in the construction of the floors of the porches outside their houses and may therefore be exposed to contaminants either dermally or through inhalation of dust.

Although the concentrations are not believed to be very high, insofar as this can be determined from the non-quantitative screening analysis employed, the presence of far higher concentrations of di- and trichlorobenzenes was established in 1990 (National Toxics Campaign Fund 1990) and further breaching of the already damaged containment liner is likely to increase the levels of exposure.

In 1999, the extent and severity of the contamination well water near the former plant was documented. Chlorinated methanes, ethanes and benzenes were found in concentrations in excess of WHO drinking water guidelines at most of the sites sampled. For example, the worst well, situated between the northern wall of the plant and the evaporation ponds contained carbon tetrachloride 1,700 times the World Health Organization (WHO) guidelines (Labunska et al. 1999). Since then, clean drinking water has been provided to these populations. However, well water is still used for washing and bathing and inhalation and dermal exposure to waterborne organochlorines can equal or exceed the intake from ingestion (McKone 1987, Wallace 1997, Moody and Chu 1995). Some risk to the inhabitants of the bustees (shanty settlements) is therefore to be expected and will probably not be eliminated until full cleanup of the groundwater resources is completed.

In 1989, Union Carbide paid $470 million in compensation for gas-related injuries and deaths (Chouhan et al. 1994) though according to non-governmental organizations involved in the court cases, many claims remain unsettled and the individual sums received are extremely small; on average $400 for injury and a maximum of $1250 for death of a family member. Legal proceedings currently underway in the U.S. and Indian courts relate to the criminal liability of Union Carbide and nine executives in place at the time of the disaster. Moreover, the issue of ongoing toxic exposure from contamination of soil and waster around the factory site has been raised as separate grounds for damages and environmental cleanup in the US case.

This report, in combination with previous works (e.g. Labunska et al. 1999, National Toxics Campaign Fund 1990), provide unequivocal evidence of a continuing risk to the local population and the environment, with the potential for these to increase, rather than decrease, over time, as degradation of the various structures in and around the plant and the continuing action of physicochemical dispersion processes lead to the further dispersion of the contaminant and stockpile inventory.

Greenpeace has recently published guidelines on the standards that will be required in the cleanup of the site, the nearby solar evaporation ponds and the groundwater resources (Stringer and Johnston 2002). Our current findings underscore the urgency of action to carry out this cleanup.

Source: Conclusions section of the 2002 Greenpeace report (pp.31–33).

Bhopal: Was the drama necessary? (2004)

Kalpana Sharma, a regarded Indian journalist, investigates why it took a hunger strike by three Bhopal survivors/activists, and an international campaign, to get the Indian government to agree that Union Carbide should clean up the factory site.

A two-page press release, issued on 23 June [2004] by the Ministry of Chemicals and Fertilizers, marked the end of a week of high drama. It stated that the Government of India had no objection to a U.S. Federal Court asking Union Carbide to clean up the mess it had left behind 20 years ago. It was also the culmination of three months of intensive campaigning by the International Campaign for Justice in Bhopal (ICJB) and Greenpeace.

The 17 March U.S. Federal Court ruling was in response to a suit filed by some of the victims of the 1984 Bhopal gas tragedy, when methyl isocyanate leaked from a plant run by Union Carbide India Limited (UCIL) killing thousands in its wake. The court ruled that the parent company, Union Carbide Corporation (UCC), should clean up the abandoned and heavily contaminated site of the now closed plant.

For this to happen, the Madhya Pradesh Government and the Centre had to state that they had no objection. What seemed on the surface to be a straightforward affair, particularly as it did not involve any costs to be borne by either Government, became a protracted affair with three people going on a fast unto death.

Was all this necessary? Given the tame manner in which the drama finally ended, it would seem not. The ICJB and Greenpeace launched their campaign first in Madhya Pradesh, urging the State Government to issue a letter of no objection. They were given the run around and told it was outside the jurisdiction of the State Government.

On 25 March, a delegation from Bhopal met the President and he expressed concern and support. After that nothing moved forward, partly because the country was by then in the election mode.

Launch of campaign

Finally, on 8 May, a campaign to petition the Government was launched. By early June, the Prime Minister, who had only recently assumed office, was inundated with over 4,000 such petitions.

On 7 June, after months of lobbying, the Madhya Pradesh Government finally sent a letter to the Secretary of the Union Ministry of Chemicals and Fertilizers, saying it had no objection if the U.S. Court ordered Union Carbide to clean up the site and that this would be "in larger public interest."

With this letter in hand, the activists then met the Union Chemicals Minister, Ram Vilas Paswan, and he promised to take action. They waited for a week and on 16 June met the Union Law Minister, H.R. Bhardwaj. The latter apparently raised the non-issue of conflict with the Bhopal Gas Leak Disaster (Processing of Claims) Act, 1985. This law had allowed the Centre to represent the claims of the Bhopal victims and finally led to an out-of-court settlement with Union Carbide amounting to $470 million. Despite the activists pointing out that this matter concerned pollution caused after the accident, the Law Ministry was unresponsive. In the meantime, several legal luminaries expressed the opinion that there was nothing in the law that need hold the Government back from issuing the letter.

By 17 June, despite more reassurances from Mr. Paswan, the activists were beginning to despair. This is when three of them decided to go on a fast. Their decision caught the attention of the media, several members of the Government and leading trade unionists.

PMO's intervention

None of this made a difference, however, until the Prime Minister's Office intervened. On its advice, the Bhopal activists again met the Law Minister on 21 June and found, to their surprise, a complete turnabout. He said he had no problem but that the Ministry of Environment and Forests had to deal with this. The latter threw the ball back at the Law Ministry. Representatives from the campaigning groups sat in all the different Ministries – Law, Environment, Chemicals – and the PMO on 22 June, waiting for some definite word. This finally came late 23 June in the form of the press release. The three broke their fast and everyone heaved a sigh of relief. But the story does not end here and there are many important lessons to be drawn.

First, the Bhopal campaigners succeeded because they had the ability to launch a campaign at different levels. There are many civil society groups without such support or such a high level of organization. Their representatives sit on dharna at various locations in State capitals and in

New Delhi and often neither the Government nor the media pays any heed to them.

Secondly, the Bhopal activists were lucky that they conducted their campaign at a time when there was a responsive Prime Minister who intervened. It is evident that without a word from his office, the matter would not have moved.

Third, the issue did not involve either the Madhya Pradesh Government or the Centre incurring any costs. They will be borne by Union Carbide according to the U.S. court's ruling. This also made the issue somewhat simpler. Yet despite this last point, it is surprising that the Bhopal campaigners had to resort to all the tricks in their bag to finally get the Government to agree. The matter should never have reached this stage and could have been settled through dialogue. The fact that the campaigners had to push things to such an extreme illustrates yet again the gap in understanding between governments and activists. The former will not respond until the latter pushes the issue to an extreme. As a result, the latter become convinced that reasonable dialogue cannot work and only extreme pressure will.

Source: Kalpana Sharma, *The Hindu,* 27 June 2004.

Via hand delivery (2004)

A statement from the Government of India to a Federal District Court judge in New York that it would have no objection if the court were to rule that Union Carbide Corporation should be required to clean up the factory site.

June 28, 2004

VIA HAND DELIVERY

United States District Judge John F. Keenan
United States District Court
Southern District of New York
500 Pearl Street,
New York, New York 10007-1312

Re : Bano et al v. Union Carbide 99 Civ. 11329 [JFK]
TO THE UNITED STATES DISTRICT COURT:

On behalf of the Union of India and as its duly authorized consular representative in the United States of America, we submit this letter in the above, referenced matter to present the official position of the sovereign government of India with regard to environmental remediation of the land and premises formerly occupied by the Union Carbide plant in Bhopal, India.

The Union of India submits that neither the Madhya Pradesh State Government or its instrumentalities nor the Union of India has any objection to any such relief for environmental remediation of the former Union Carbide plant premises in Bhopal being ordered or directed by a competent court or tribunal of the United States. Further, the Union of India and the Madhya Pradesh State Government and their respective instrumentalities will cooperate with any such relief as and when issued by the United States District Court. The Union of India will monitor and supervise such environmental remediation including decommissioning of plant and machinery, remediation / disposal of contaminated soil and appropriate disposal of toxic chemicals and wastes on the plant site by Union Carbide in order to ensure that it is undertaken in compliance with the norms parameters laid down by a specific organization of the Government of India, the Central Pollution Control Board, for that purpose.

Union Carbide will also be held responsible for any loss/damages caused to life or property in the process of remediation and disposal. Pursuant to the "polluter pays" principle recognized by both the United States and India, Union Carbide should bear all of the financial burden and cost for the purpose of environmental clean up and remediation. The Union of India and the State Government of Madhya Pradesh shall not bear any financial burden for this purpose.

Notwithstanding the foregoing, nothing in this official statement on behalf of the Union of India may be construed or read, by implication or otherwise, as an intention to submit either the Union of India or the Madhya Pradesh Government to the jurisdiction of the United States. The Union of India and the Government of Madhya Pradesh are entitled to sovereign immunity under international law and do not waive those immunities by this submission.

In addition, nothing in this submission should be construed, by implication or otherwise, to convey any authority to plaintiff in the above matter to assert or pursue claims on behalf of the Union of India or State Government of Madhya Pradesh nor shall the plaintiffs in the above referenced matter be entitled, by virtue of this submission, to assert or pursue any claims against either the Union of India or the Madhya Pradesh Government in the litigation or before the US District Court.

Finally, it is the official position of the Union of India that the previous settlement of claims concerning the 1984 Bhopal Gas Disaster between

Union Carbide and Union of India has no legal bearing on or relation whatsoever to the environmental contamination issue raised in the case at bar. Nothing in this submission should be construed, by implication or otherwise, as an intention to reopen or question the validity of that previous settlement.

Accordingly, the Union of India hereby formally urges the US District Court to order such relief, as required by the US Court of Appeals Second Circuit in this matter.

Respectfully submitted

Consul General of India
Consulate General of India
3 East, 64th Street
New York, New York 10021-7097

Source: www.bhopal.net

9

Healing

Failure and innovation in treatment

IN THE HOSPITALS of Bhopal, the effects of both acute (gas leak) and chronic (water contamination) chemical exposure have been made nearly invisible by a simple omission. The suppression of research and data about the combination of symptoms that characterize exposure to MIC and its component chemicals make it almost impossible to diagnose syndromes: only symptoms. This distinction is crucial for two reasons: first, classification of disease and exposure directly affects the patient's rights to treatment and compensation (as well as the larger statistical understanding of the magnitude of the problem); and second, because this guarantees that no one will ever be cured – they will only be treated symptomatically. The stark truth is that the continued sickness of the gas and water affected has been a continuing boon for the doctors, hospitals and pharmaceutical companies in Bhopal.

Most gas-affected people who had the resources to do so were treated privately, often outside of Bhopal or even outside of India, for their gas exposure. However, it was the poorest communities that were hardest hit – and the poor did not have the resources to get treatment externally. And unfortunately, they have by and large not benefited from the public hospitals and clinics that have been constructed in Bhopal to serve them, such as the Bhopal Memorial Hospital Trust (BMHT) established with money from the sale of Union Carbide's Indian assets. BMHT's distance from the affected communities, limited working hours (closed in the evenings when the poor can make time), emphasis on expensive diagnostic equipment and unnecessary amenities (such as swimming pools for its staff), refusal to recognize water contamination or next generation effects as gas related, and failure to develop a treatment protocol, make it clear that its interest in effectively and permanently healing the gas affected is limited. Government sponsored treatment facilities manifest similar problems.

It is a serious medical challenge to evolve treatment methods for a situation as complex and understudied as the effects of gas exposure – particularly as the reverberations of this exposure begin to be felt in the next generation. This chapter outlines the problems that have plagued the effort to heal the people, and the city, from the injury of the 1984 leak. It also discusses one of the proactive responses from Bhopal survivors, local activists, and a handful of medical professionals and social workers: the creation of the Sambhavna Clinic, a free medical treatment center where modern medicine, Ayurveda and yoga are combined with community outreach and health education in an attempt to actually cure the gas affected.

Medical crime (2004)

Satinath Sarangi charts the troubled history of attempts to address the medical crisis in Bhopal, beginning with the thiosulphate controversy in the disaster's immediate aftermath, and ending with the effects of groundwater contamination.

On the midnight of 24 June 1985, a colleague and I were at the Jana Swasthya Kendra, a clinic that gave sodium thiosulphate injections to people from the communities closest to the Union Carbide factory in Bhopal. The kendra was supported by four local organizations of survivors and workers. Run by volunteer-doctors from Calcutta, Bombay and other parts of the country, the kendra also monitored the effect of this drug on the many symptoms the exposed people suffered. In just twenty days of its running, the kendra had administered more injections than all government hospitals had done in the past six months.

As managers, we stayed at the clinic at night to get it running by early morning. On this particular night, a dozen armed policemen entered the clinic, forced us into two separate jeeps and took us away to separate police stations where we were locked up till morning and sent to jail the next day. Several of the volunteer-doctors and activist survivors were also arrested and jailed. The charges cooked up against us were those of attempting to murder government officials and committing other serious offences. As we found out a few days later, after sending us away, the rest of the posse took away over 1200 medical folders that contained records of the beneficial effect of sodium thiosulphate injections. These records were never returned.

Sodium thiosulphate injection was literally a life and death issue in the immediate aftermath of the disaster. Dr Max Daunderer from Germany and local forensic specialist Dr Heeresh Chandra both found that the drug, when administered intravenously, led to excretion of increased levels of thiocyanate in urine. Double blind clinical trials carried out by the Indian Council of Medical Research from 1985 to 1987 confirmed the efficacy of this drug in relieving exposure induced symptoms and causing detoxification of the body.[1]

Ten days after the disaster, Union Carbide Corporation's medical director first supported mass administration of thiosulphate and, in another telex message three days later, forbade it. Soon after, Union Carbide's ally in the state bureaucracy, Director of Health Services, Dr M.N. Nagu sent a circular to all doctors in the city warning them that they would be held responsible for any untoward consequences of thiosulphate administration. In the prevailing situation of medical uncertainty this circular effectively stopped any administration of thiosulphate. Interestingly, given that no adverse impact of administration of thiosulphate was reported in literature, the circular had no medical basis.

In April 1991, my friend T. R. Chouhan, a former MIC plant operator in Union Carbide's Bhopal factory, and I met with Joseph Geoghen, then Vice President of Union Carbide Corporation, USA in his corporate office in New York city. Earlier, in 1989, the Chairman and CEO of Union Carbide, Robert Kennedy had extended an invitation to the group of three Bhopal survivors whom I had accompanied on a campaign tour. Carbide had got all four of us housed in Houston county jail on charges of criminal trespass. In 1991, we decided to talk about Carbide's toxic trespass into the bodies of the people of Bhopal.

Goeghen met us with two lawyers who vetted each statement he made. They whispered their advice into his ears from either side as we insisted on recording our discussion on tape. I described the deteriorating health condition of the survivors. I told him how doctors were unable to treat people and no medicines seem to be working for the exposure related illnesses. We pointed out that what doctors in Bhopal needed was information about the health effects of the leaked gases. This would help in developing treatment protocols and in identifying specific areas of research, we pleaded. This wouldn't even cost money, we assured him.

I mentioned the names of the different laboratories where Union Carbide had carried out tests with methyl isocyanate, the major component of the leaked gases. Goeghen did not seem to recognize any of them. In particular I mentioned the research carried out at Carnegie Mellon Institute in Pittsburgh where the corporation had documented the effect of MIC on living systems in the '60s and '70s. Goeghen refused to get into specifics. He told us we could get Material Safety Data Sheets published by official agencies on MIC and other chemicals. We informed him that we had already looked them up and did not find the information we needed.

Goeghen insisted that all information with Union Carbide had been passed on to the Indian government. Chouhan pointed out that at the very least Union Carbide could disclose the medical records of the workers who were subjected to routine examinations in the factory but whose results were withheld from them. Goeghen would not reply to that. He indicated he was pressed for time. One of his lawyers had a flight to catch.

In May 2001, about three months after Union Carbide merged with Dow, as part of our negotiations I presented a brief note titled "Two humanitarian things the Dow Chemical Company, USA can immediately do to help survivors of Bhopal, India" to the Managing Director of Dow India in Mumbai. The first was to release the unpublished medical information. In November 2001, the Managing Director wrote that they were sending us an inventory of published medical research and that not only had they not found any unpublished research, they were unable to locate anyone in the UCC organization who knew of such research.

In November 1991 about a dozen of us carried out a survey in three gas affected communities to find how the health damages suffered by people had been assessed by the medicolegal authorities.[2] We found that the procedure of injury evaluation formulated by the state government's Directorate of Claims grossly underestimated the range and degree of injuries caused by toxic exposure. We found that only 10% of the claimants in the most severely affected community had been given the Pulmonary Function Test. Only 18% had had their eyes tested. While ICMR studies were reporting gas-exposed women having an abortion rate five times that of unexposed women, even five years after the disaster,[3] only 11% of women claimants had been examined by a gynaecologist.

Leaving room for procedural errors in an exercise of such large magnitude, the number of claimants from the 36 municipal wards matched well with the number of residents. Yet the results of categorization of claimants were completely at variance with the epidemiological research on the exposed population. While ICMR found that immediately after the disaster 99% of the population had breathlessness, 86% had eye problems and 92% had loss of appetite,[4] the Directorate's medical evaluation reported that 42% of the claimants had suffered no injury at all.

The Directorate declared that 52% of the claimants had only temporary injury. This while ICMR researchers were finding that "one to three months post-exposure, a majority of the already affected population continued to suffer from breathlessness, cough, chest pain etc...." According to the Directorate, only two persons from the two most severely affected municipal wards had suffered injury in the "most severe" category. In fact, the findings of the Directorate were so scandalous that the authorities in charge of fixing compensation decided that those categorized as having not suffered any injury at all would be considered at par with those who were considered to have suffered temporary injury.

In September-October 2002, I was part of another survey in Jai Prakash Nagar, a settlement right opposite the Union Carbide factory.[5] In this house-to-house survey we found that 91% claimants in this area – bore the worst brunt of the toxic attack – had received only Rs.25000 as compensation and most received their compensation eight to ten years after they registered their complaint.

What role did medical professionals play in sustaining an irrational and unscientific system designed to down play the damages caused to a majority of the people?

On 16 July 1988, Dr N.R. Bhandari, a Professor of Paediatrics, presented the findings of his team's study on children born after the disaster to exposed mothers before the Scientific Advisory Committee.[6] As compared to children of unexposed mothers, these children were found to have delayed physical and mental development and lower values for anthropometric parameters such as height and mid-arm circumference. According to the minutes of the meeting, the work of Dr. Bhandari and his team was appreciated by senior scientists and he was asked to continue the study till the children were 14. The 16 May 1989 meeting of the Project Advisory Committee recommended the continuation of the study and further recommended that the children's sexual development and immunological functions be also studied. The same committee in September 1990 reiterated that the children must be studied till they attain puberty.[7] Despite the positive and significant findings regarding teratogenic effects of the toxic exposure, and in the face of opposition from its Principal Investigator, the study was wound up in June 1991 following directions from the headquarters of the Indian Council of Medical Research.

Ten years later (in May 2001), I was part of a team that carried out anthropometric measurements of teenagers born to exposed mothers. We found that compared to age-matched teenagers whose mothers were unexposed, male children born to exposed mothers were shorter, thinner, lighter and had smaller cranial and mid-arm circumference. The study was published in an international medical journal.[8]

Why was the study that showed injuries in the next generation of survivors prematurely terminated?

In June 1990, with help from a group of young doctors we interviewed 500 gas-affected patients in two government hospitals and collated information from their medical prescriptions.[9] We found that 53% of the patients had been prescribed medicines that were banned in several countries and were considered fit for banning by the Indian government. 40% of the medicines prescribed were irrational and hazardous. In 1994, Drs Rajiv Bhatia and Gianni Tognoni of the International Medical Commission on Bhopal (a 15 member international team of voluntary medical professionals), found irresponsible and indiscriminate use of antibiotics and corticosteroids among the gas exposed population.[10]

In October 1996, we collected data on medicines sold through 50 drug stores in the gas-affected area. We also recorded the medicines listed on prescriptions brought by 200 customers to 25 of these shops. We found that among the most sold 391 medicines, 46% were harmful, hazardous or useless and that 52% of the drug market was controlled by multinational pharmaceutical corporations. One out of every three prescriptions by a qualified medical practitioner was found to be irrational.[11]

From June to November 1998, we collected information on the treatment given at one of the community clinics run by the Bhopal Memorial Hospital Trust from 474 gas affected patients and their health books. Analysis of the data showed that drugs prescribed were not targeted to the organ system damage but towards short-term symptomatic relief. Data also showed the use of high potency systemic corticosteroids that could increase susceptibility to tuberculosis.

In 1997, we began documenting the beneficial effects of simple yoga postures and a few *pranayama* breathing exercises on people with respiratory disorders caused by their exposure to Union Carbide's gases.

The participants were initially trained in the exercises for 15 days by two yoga therapists at the clinic who also monitored the condition of their lungs through spirometry and physical examination for six months. The study reported significant and sustained improvement in lung function parameters for all participants. Half of the people in the study could do away with the bronchodilators they had been so dependent upon ever since the disaster.

After the study was published, in 1998,[12] the two yoga therapists sent reprints of the paper to senior officials of the state government's Bhopal gas tragedy relief and rehabilitation department and the Bhopal Memorial Hospital Trust. In the covering note, the Sambhavna therapists sought their opinion on the paper and on introducing yoga and pranayama in the hospitals and clinics run by them. When they did not receive any reply for one month, they sent reminders. The government official sent appreciation for the paper and regretted that yoga could not be introduced in government hospitals because it was not possible to find so many instructors. At a stretch, this excuse could have been accepted in any other city but Bhopal has a surfeit of yoga instructors, thanks to the efforts of Dr K.M. Ganguly who has helped train thousands of yoga teachers.

In February 1985, a team of four doctors observed that women who had been pregnant at the time of the disaster had reported spontaneous abortions, stillbirths and menstrual disturbances.[13] Clinic-based information generated by independent physicians in February-March 1985 indicated large number of menstrual and gynaecological disorders.[14] Another study initiated by a survivors' organization showed that 50% of the women who were clinically examined had persistent gynaecological symptoms with excessive vaginal secretion [leucorrhoea] being the commonest.[15] In the same month, a study carried out by an independent team showed that among women in the age group of 15-45 years there was a significant alteration in the menstrual cycle, excessive bleeding during menstruation, and dysmenorrhoea.[16]

From March 1985, when data indicated the presence of exposure induced gynaecological diseases, persistent attempts were made by survivors' organizations and their supporters to include the gynaecological impact of toxic exposure in the proposed research to be carried out by the Indian Council of Medical Research. However, these attempts were unsuccessful and none among the 24 research projects identified by the ICMR concerned themselves with documenting the gynaecological impact of the disaster unless it was related to fertility.

In the June 1999 issue of *Meri Saheli*, a Hindi monthly magazine, a copywriter trying to raise funds for the Sambhavna Clinic published an advertisement on the health situation in Bhopal. The advertisement focused on the continuing gynaecological impact of the gas five years after the disaster. Earlier in February the same year, the copywriter had visited gas affected communities and interviewed teenage women. The publication of the advertisement in a local newspaper was followed by threats of police action and worse. The Chief Medical Officer announced to the media that he would lodge a criminal complaint against me for spreading alarm.[17] The Principal Secretary of the Department of Bhopal Gas Tragedy Relief and Rehabilitation of the state government told another newspaper that he had asked officials to explore legal possibilities.

In July 2001, the state government proposed the setting up of an Institute of Life Sciences at Bhopal at an estimated cost of Rs.200 crore. The proposal sent to the Ministry of Chemicals and Petrochemicals, the nodal ministry for issues concerning the disaster in Bhopal, stated that the institute was "mainly to study the deleterious effects of MIC which were seen in humans, plants and animals (following the gas tragedy of December 1984) with a view to finding genetic solutions."[18] "The proposed Institute of Life Sciences," the proposal went on to say, "would be an excellent Centre to provide the infrastructure and manpower to find the genetic solution to reduce the sufferings using the various modern day molecular biology and biodiversity techniques such as RFLP, RAPD, PCR, DNA fingerprinting and gene therapy."

Commenting on the state government's proposal, Dr P.M. Bhargava founder Director of the Centre for Cellular and Molecular Biology, Hyderabad, wrote, "It is clear that whosoever has written the note has virtually no understanding of modern molecular biology, including genetics. Scientific jargon has been used in the note without understanding what the terms mean and merely to lend some credibility to the idea of setting up a Institute of Life Sciences amongst those who have no idea of modern biology or even the problems of the Bhopal gas tragedy victims. To try to set up such an institute with the money that was meant to be a compensation to the gas tragedy victims for all the pain, suffering and loss inflicted on them is nothing short of an arrogant and dishonest attempt to hoodwink and divert the funds that were not meant for the purpose for which they are being attempted to be diverted. Indeed, no genetic solutions are required to take care of the present problems of the gas tragedy victims."

In April 1990, after several people living in the vicinity of the abandoned factory had complained about the foul smell and taste of the water in their hand pumps, we sent samples of ground water and soil from the vicinity of the Union Carbide factory to the Citizen's Environmental Laboratory in Boston, USA. The test reports showed the presence of chemicals that damage the lungs, kidneys, liver and produce cancer. These reports were presented at Union Carbide's annual general meeting the same month.

Reports of tests carried out by the state government's Public Health Engineering Department in 1991 and 1996 confirmed severe contamination of ground water in several communities adjacent to the factory. In 1999, Greenpeace scientists tested 22 samples of groundwater from in and around the factory site. They found heavy concentrations of chlorinated benzenes, carbon tetrachloride, trichloroethene and other chemicals that cause various cancers and damage the liver, kidneys, brain, immune system and other organs. In 2003/2004, samples from 13 locations tested by the Madhya Pradesh Pollution Control Board showed the presence of lindane, benzene hexachloride and other hazardous chemicals.

Despite the alarming test reports from government agencies from 1990 to 2003, the government continued to deny the contamination of ground water. The minister for Bhopal Gas Tragedy Relief and Rehabilitation – who ironically was a medical professional himself – announced to the media that the ground water was safe to drink. Another minister drank a glass of the hand pump water before media persons, *a la* Norman Borlaug, in a bid to make the denial stronger

Meanwhile, studies by Sambhavna showed high prevalence of severe anaemia, known to be caused by ingestion of trichlorobenzene present in ample amounts in the community hand pump water, among the resident population. That every second person in the community was suffering from symptoms such as abdominal pain, skin disorders, giddiness, chest pain, headache and fever. That among the teenage females in the age group of 13 to 15, 43% had not begun their periods.

I guess, I am now expected to make my point, elaborate on the meaning of the stories, draw upon their interconnectedness and present a framework that holds them together. That would however, be straying away from why I really wanted to tell these stories. Why I really began telling these stories was to move you, dear reader, to action. Twenty years is much too long and we have had a lot of words. No more interpretations, no more words – the point is to stop the medical disaster in Bhopal.

Footnotes:

1. S. Sriramachari, "The Bhopal Gas Tragedy: An Environmental Disaster," *Current Science* 86 (7), 10 April 2004.
2. "Compensation Disbursement: problems and possibilities." Bhopal Group for Information and Action, Bhopal, January 1992.

3. "Health Effects of the Toxic Gas Leak from the Union Carbide Methyl Isocyanate Plant in Bhopal." Indian Council of Medical Research, Ansari Nagar, New Delhi, May 2004, p.50.
4. Ibid.
5. Minutes of the meeting of the Scientific Advisory Committee for Bhopal Gas Disaster Research Centre held on 16.8.88 at Bhopal.
6. Letter to the Director General, Indian Council of Medical Research from Professor N.R. Bhandari, Chief Investigator, ICMR Project, 1 December 1990.
7. Nishant Ranjan, Satinath Sarangi, V.T. Padmanabhan, Steve Holleran, Rajasekhar Ramakrishnan, Daya R. Varma, "Methyl Isocyanate Exposure and Growth Patterns of Adolescents in Bhopal," *Journal of American Medical Association* 290(14), 8 October 2003.
8. "Evaluation of Some Aspects of Medical Treatment of Bhopal Gas Victims." Bhopal Group for Information and Action and Socially Active Medicos, Indore, August 1990.
9. Rajiv Bhatia and Gianni Tognoni, "Pharmaceutical Use in the Victims of the Carbide Gas Exposure," *International Perspectives in Public Health*, Volumes 11 & 12, 1996.
10. "The Bhopal Gas Tragedy: 1984" – A report from the Sambhavna Trust, Bhopal, November 1998.
11. A. Gupta, S. Durgavanshi and I Eckerman, "Effects of yoga practices for respiratory disorders related to the Union Carbide gas disaster in 1984." XVI World Congress of Asthma, Buenos Aires, Argentina, 17–20 October 1999, pp.83–87.
12. "Medical Relief and Research in Bhopal: the realities and recommendations." Medico Friends Circle, February 1985.
13. Rani Bang, "Effects of the Bhopal Disaster on Women's Health: an epidemic of gynaecological diseases." Mimeograph, 1985.
14. "Medical Survey on Bhopal Gas Victims between 104 and 109 Days after Exposure to MIC Gas." Nagarik Rahat aur Punarvas Samiti, March 1985.
15. "The Bhopal Disaster Aftermath: an epidemiological and sociomedical survey." Medico Friends Circle, October 1985.
16. "Storm in Bhopal Over Ad on Gas-hit Girls," *The Times of India*, 1 June 1999.
17. Letter from Alka Sirohi, Principal Secretary, Government of Madhya Pradesh, Department of Bhopal Gas Tragedy Relief and Rehabilitation, to N.C. Gupta, Joint Secretary (Bhopal), Ministry of Chemicals and Petrochemicals, 9 July 2001.
18. Personal communication from Dr P.M. Bhargava; 28 October 2003.
19. "The Bhopal Legacy." Greenpeace Research Laboratories, University of Exeter, November 1999.

Source: From *Seminar 544: Elusive Justice*, December 2004, New Delhi, India.

Health infrastructure for the Bhopal gas victims (1996)

This report, researched in 1994 by M. Verweij, M.D., S.C. Mohapatra, M.D. and R. Bhatia M.D., details the limitations of the health infrastructure existing in Bhopal, and outlines a model for the kind of community based health system still needed.

Introduction

The immediate aftermath of the Bhopal gas disaster tremendously overloaded the health care system. In the course of a few days approximately 180,000 outpatients and 11,000 inpatients crowded a 1,000-bed hospital. Many died even before reaching health facilities. Health facilities are usually planned on the basis of the morbidity and expected medical consumption within a certain population. It is very understandable that the available health care facilities had difficulty coping with a disaster of this magnitude.

Today more than nine years after the disaster, the gas victims still complain that they have difficulty finding adequate health care. The complaints include having to wait in long queues, receiving inadequate and irrational expensive drug prescriptions, and not being able to find a satisfactory treatment for their chronic symptoms. The clinical symptomatology of the Bhopal syndrome will be discussed in other chapters of this International Medical Commission Bhopal Report. It is clear, however, from clinical assessment, that there is long term multi-organ morbidity and that the health needs of the Bhopal community will remain increased. Therefore the health infrastructure will need to be adapted to these higher health needs within in the limitation of resources in the Indian context.

In table 1, some general demographic information is presented in relation to the population of Bhopal

Table 1. Demographic information

Population of Bhopal (1991 census)	1.06 million
Estimated population of Bhopal 1995	1.22 million
Percentage of population under 5	18%
Percentage of population under 15	37%
Population in 36 gas affected wards	701,470
Average population per ward	19,500
Total number of dispensaries	36
Total number of hospital beds for the city	2,100

Source: Report mentioned in note 2 by the Department of Preventive and Social Medicine of the Gandhi Medical College. The survey data in the report states that in the slum areas more than 50 percent of the population is under 15 years of age.

The 36 gas affected wards (a ward being an administrative area within the city) out of the 56 wards in Bhopal is the part of the city served by the health department of the Ministry of Gas Relief. Due to migration, however, the current population includes exposed and unexposed people.

A subgroup of the International Medical Commission on Bhopal (IMCB) for "Community needs and health infrastructure" developed the following methodology to assess the required health infrastructure. Firstly the available health care was assessed through meeting with government officials, private organizations, private practitioners and the community representatives. Secondly, the complaints concerning the health care were analyzed. Thirdly, conclusions were drawn and recommendations made...

Health Facilities Required

In the discussions with the community and the government, the issue of the number of available hospital beds was not a major one. There are approximately 1.72 hospital beds per 1000 in Bhopal and this is more than sufficient under the present circumstance as many health services are delivered on an outpatient basis. The 1993 World Bank report, "Investing in Health" states that the average number of beds per 1000 for the whole Madhya Pradesh State is 0.4/1000. The same report recommends that 1.0 hospital bed per 1,000 is sufficient to establish basic hospital services in developing countries.

Taking into consideration that most of the resources for relief of the health problems of the gas victims have gone into hospital-based, institutional infrastructure, the emphasis should in the future be towards outpatient and community services. This coincides with the fact that most of the specific gas-related health problems of the victims can be dealt with on an outpatient basis.

From government sources it was found that the total number of OPD consultations in the gas relief hospitals and clinics was approximately 1.43 million in 1992. This number, of course, excludes the visits made to the private practitioners. This means that on average a person uses a government health facility 2.04 times per annum. These numbers of outpatient consultations are so enormous that it is not surprising that the existing government facilities are overstretched. If all patients were to go to the primary level – the 14 dispensaries – it would mean they would have to handle approximately 100,000 consultations per dispensary per year.

Effectively we can conclude that the entire bottom tier of the health care pyramid is more or less nonexistent which in any situation would cause very serious problems for adequate health-care delivery. This results in doctors having very little time for each patient and continuously feeling overburdened. Likewise it creates consumer dissatisfaction with the patients.

Outline of a Possible Model for Health Infrastructure

The most effective way of organizing the health care system would be to adopt a pyramid structure with four tiers. This means that an integrated system of health care is provided with the higher level providing backup and support to the lower levels. In the present situation the top two tiers are available and existing but are overstretched due to the fact that the bottom two levels, the primary and community level, are underdeveloped or non existent.

At present the 14 available primary level dispensaries cater for an average population of 50,000 people. This is too large a catchment area for a primary health facility to cover. Interaction with the community is not feasible with this magnitude of catchment population. Therefore a tier at the community level would need to be added with facilities having a catchment population of not more than approximately 5,000 people. This would imply that if the health consumption of the people were to remain the same, these community health facilities would still be handling more than 10,000 consultations per annum.

The community health level could be staffed by paramedical staff. There are several existing cadres in India e.g., multi purpose worker (MPW) community health volunteer (CHV) and trained traditional birth attendant (TTBA) that could be adopted for the community level. Existing training programs within the government and non government sector are available. From the primary level upwards the staffing will include medical doctors.

Assuming that the community has faith in the services provided, not more than 25 percent would need to filter through from the community to the primary level, and not more than 15 percent from the primary levels upwards. The proposed health pyramid is given in table 2.

This model is based on the situation in January 1994. Since then more hospital facilities have been established. These would have to be assessed in order to be included in this model. The theoretical base of the model, however, remains relevant. It implies that far less medical doctors would be needed for the primary and community levels. This would set the conditions for more time per patient at the higher levels of the pyramid.

It should be noted at this point that community health workers should be well trained and committed. Otherwise people will lose faith and bypass the community level of the pyramid structure. As already mentioned, an urban context offers many possibilities for medical shopping. Some form of

Table 2*: Proposed Health Pyramid

Level of care	Facility available	Estimated patients	Patients/ Facility
TERITIARY (1 unit)	MIC-ward, Hamidia Hospital Gandhi Medical College		
SECONDARY (3+1 UNIT)	Jawahar Lal Nehru Hospital Khan Shakir Ali Khan Hospital Master Lal Singh Hospital Bhopal Eye Hospital	53, 500	
PRIMARY (14 units)	Rukma Bai Clinic Ibrahimganj Clinic 8 Dispensaries 4 Red Cross Clinics Dispensaries	357, 500	25, 500
COMMUNITY (140 units)	Community Health Unit	1,430, 000	10, 000

table No. 5 in the original

"gate keeping" system could be introduced between the tiers; for instance preferential treatment for referred patients and the use of colored referral cards.

Financial Aspects of the Proposed Model

The model described above with much more emphasis on community health care, but at the same time offering basic curative services in hospitals, is comparable with the model described by the World Bank as the basic health package. In its report "Investing in Health," (1993) the World Bank estimates the cost for this package at $8 per person per year in low income countries. For the population of the gas affected wards this would be $5.6 million per annum. In comparison to the settlement sum paid by Union Carbide this is a very modest expenditure. Theoretically the settlement sum could have provided the gas victims with an adequate level of health care for close to seventy years.

Conclusions and Recommendations

First it should be noted that recommendations for a primary health care approach are not limited to the gas victims. This approach is relevant to all the people of Bhopal, especially in the slum areas. Due to the specific health needs of the gas victims, implementation should be given priority in affected areas. In the longer term, however, it is advisable to reintegrate the vertical health-care system established for the gas victims into the general health-care system of the state. A vertical system is vulnerable and has the risk of coordination problems and duplication of services.

I. In order to establish an adequate health care system, the community level and the primary level of the health care pyramid should be developed. There is an urgent need for a primary health-care approach in the gas affected wards, which is in line with the official health policy of the government of India. It remains a challenge to establish such a program in an urban setting. This could be implemented through three scenarios:

Scenario A: that the Ministry of Gas Relief urgently implements its plans to decentralize the existing health care as described in its annexure 5 of the updated December 1993 report (see note 1). Reference is made to the detailed plans for a possible health-care model for slum areas presented in the report of the Department of Preventive and Social Medicine (see note 3)

Scenario B: that the community establishes community health-care services itself. This will need initial support from professionals and other voluntary health agencies, but should be able to run on its own after the initial phase. As several community groups have an existing organizational infrastructure, these could be used to find suitable people to receive further health training.

Scenario C: that a model is established by which the community groups and government cooperate together to establish and reinforce the primary and community levels of the system. This would imply community participation in the planning and implementation of the services. Considering the limitations of the available resources, the commission feels that this would be the most effective scenario.

At the community level, health units (whether formal or informal) should not serve more than 5,000 people. It would be possible to run this level of service with paramedical cadres. The existing dispensaries (14) could be upgraded to be the first referral center at a primary level. This level would need the services of medical doctors, serving an average catchment population of 50,000 people. For the implementation of this kind of program it is recommended that other urban health projects in India be studied and used as reference models. A facilitator will be needed for the actual execution.

That the mentioned primary and community levels of the health-care pyramid should be developed according to the principles of the Primary Health Care approach as defined by WHO. This means that emphasis should not only be on curative medicine. Very important are the sanitary measures in the urban areas, health education and nutrition, mother and child care, maternity services, rehabilitation and physiotherapy (breathing exercises) for the gas victims and family welfare services. These aspects are especially relevant for the gas victims as additional morbidity can worsen their underlying problems due to the MIC-exposure. It is also recommended that, especially for the sake of the affected gas victims with pulmonary disease, an ongoing monitoring of the air pollution through the use of coal and cow dung in stoves is established.

Authors:

Marinus Verweij, MD DBA, the Netherlands, commissioner of the International Medical Commission on Bhopal (IMCB). Representative of the World Council of Churches. Senior Consultant with the National Board for Hospital Facilities of the Netherlands.

S C Mohapatra, MD, MPHC (UK), India, member of the national organizing committee for IMCB. Reader in the Dept. of Preventive and Social Medicine, Banaras Hindu University.

Rajiv Bhatia, MD, USA. Member of the national organizing committee for IMCB. Physician in Berkeley, California, USA

Notes:

1) From the report of the Ministry of Gas Relief, "Medical Rehabilitation of Gas Victims, an Update," December 1993.
2) The public expenditure on health in India is approximately 2 percent of the GNP. The expenditure on health in the private sector is estimated at approximately 4 percent of the GNP.
3) Exactly the same complaints are given by the slum dwellers in Bhopal in a general report by the Department of Preventive and Social Medicine of the Gandhi Medical College on health-care needs of slum dwellers in Bhopal (1993).

Source: *International Perspectives in Public Health*, Vol. 11–12, 1996.

A brief note on prescriptions collected from Bhopal Medical Hospital Trust (2000)

Because there is no treatment protocol or understanding of the combinations of symptoms that result from gas exposure, much of the treatment and many of the prescriptions at the Bhopal Memorial Hospital Trust are ineffective or harmful.

Dr Atanu Sarkar reports.
Date: 17 to 19 July 2000

The hospital and the five clinics run by Bhopal Memorial Hospital Trust were built in order to provide long awaited quality medical care to the gas victims. The wide publicity around the BMHT hospital and clinic created hope and expectation among the gas victims who had been denied basic medicare since the disaster in 1984.

The prescriptions were collected by Sambhavna clinic field workers during their visit in 1998-99 and photocopies made for documentation.

The present study is based on the photocopy of the prescriptions collected from patients undergoing treatment at BMHT, by Sambhavna Trust, Bhopal. Dr. Rajiv Bhatia of the International Medical Commission on Bhopal (IMCB) analyzed the same prescriptions and presented the following facts:

- Proportion of visits where health workers and BMHT recorded the specified symptoms and symptom categories
- Co-relations between the symptoms recorded in the health worker's interview, and symptoms written on the physicians prescriptions (first recorded visit)
- Frequency of prescriptions for the most common classes of drugs for all recorded visits
- Odd ratios for prescriptions of drugs in category for selected symptoms

The present study has looked into: 1) the nature of reported clinical examinations; 2) the appropriateness of drug prescribed; 3) the nature of dosage of appropriate drugs prescribed.

Prescriptions had been collected from Sambhavna Trust, Bhopal. Sambhavna Trust collected prescriptions of 400 individuals. Total 101 prescriptions of the patients have been selected for study and their total 380 visits have been checked. Analyses are broadly shown in three tables. Although the sample size was small when compared with the gas affected population, it gives a clear idea about general trends and guide for further course of action.

Table 1		
No description of symptoms or just few words	Mixed i.e. descriptions of symptoms in few words along with description of signs in few words	Proper description of sign and symptoms at least to diagnose the disease
252 (66.3%)	63 (16.6%)	65 (17.1%)
Total 380 visits		

Table 1 illustrates signs and symptoms mentioned in the prescriptions. There are several prescriptions without any mention of symptoms (patient's descriptions of his / her suffering/s), or with just one or two words like "pain" – without any mention of its location or intensity – and "cough" – without any mention of whether it was dry or productive – or few descriptions of signs (doctor's observation/s like – pulse rate or blood pressure only) and very few proper descriptions on signs and symptoms. Proper descriptions are very important for further follow-up and also for other physicians who may see the patients. Treatments of the patients are based on these kinds of improper examinations. In table 1, 380 visits are divided into three columns, 1) either no description of symptoms or just a few words; 2) mixed i.e. descriptions of symptoms in a few words along with description of signs in a few words; 3) proper description of signs and symptoms.

Table 2			
Harmful (may or may not be associated with proper selection of drug/s)	Useless (may or may not be associated with proper selection of drug/s)	Both Harmful & Useless (may or may not be associated with proper selection of drug/s)	Neither harmful nor useless (proper selection of drug/s)
100 (26.3%)	184 (48.5%)	29 (7.6%)	67 (17.6%)
Total 380 visits			

Table 2 illustrates the extent of irrational therapy. There are several prescriptions where drugs have been prescribed without any justification (according to any standard pharmacology text) and may harm the human body: Corex cough suppressants containing opium derivative, alprazolam as anxiolytic, benzodiazepine as anxiolytic, or combination of Corex and alprazolam, or single drug anti TB therapy, overdose of anti helminthic drugs and so on. A combination of Corex and anxiolytic may cause serious effects on the brain resulting in disorientation, drowsiness etc. Overdose of anti-helminthic drugs may cause abdominal discomfort. Single TB dose results in resistant bacteria. Several prescriptions list drugs that have little or no therapeutic value. For example, multi vitamin capsules/tabs for a few days (without mention of symptoms/signs of specific vitamin deficiency), iron capsules for few days, unnecessary antimicrobials (for mere symptoms of common cold, pain abdomen and so on) below the minimum recommended dose according to age and body weight of the patients. Some of this may result again in resistance of bacteria to antibiotics. Combination of antacid and H2-receptor blocker (ranitidine or famotidine) for just abdominal pain – abdominal pain might be due to any cause – will not give any extra benefit, but rather cause unnecessary expenditure, and so on. There are some prescriptions where both harmful and useless drugs are prescribed. It is worth noting that the aforementioned categories of prescriptions are not necessarily containing only harmful, useless or both kinds of drugs. These may or may not be associated with appropriate selection of drug/s. There are very few prescriptions where neither harmful nor useless drugs have been prescribed and where proper drugs been recommended.

Table3	
Improper dose	**Proper dose**
92 (45.3%)	111 (54.7%)
Total 203 visits	

Table 3 illustrates prescribed doses of properly selected drugs. There are a number of prescriptions where right drugs are prescribed but with wrong doses like frequency of intake per day, duration of the course and strength of each intake.

Notes:

- There are two cases where anti-TB drugs are prescribed without investigations and doses did not follow any WHO or government guidelines. There are two TB cases where proper investigation and treatment protocols have been followed.
- There are 22 visits where alprazolam, and one visit where benzodiazepine, have been prescribed without any justification.
- There are several prescriptions where concerned doctors wrote "*No sensitivity to any drugs,*" perhaps to save themselves from any possible legal suit from patient/s. This is ridiculous because on the one hand the doctors are not willing to pay more attention to examine the patients, writing signs and symptoms in the prescription sheet. On the other hand, they showed more concern to avoid possible legal action on account of drug reactions.

- Conclusion: The brief analyses of prescriptions collected from BMHT have revealed that concerned doctors are not careful about clinical examination and recording. In most of the cases, harmful and useless drugs have been prescribed. Moreover, nearly half of the rightly chosen drugs were not given following a proper dosage schedule. This study throws light on outcome of knowledge, attitude of doctors and hospital authority. It is necessary to trace back to doctor's level (attitude and practice) and authority level to corroborate these findings.

(Dr Sarkar, MBBS, holds a Masters in Community Health, JNU-New Delhi, and is a Doctoral Fellow, JNU-New Delhi; he is also Program Officer, The Catholic Health Association of India (CHAI) and Coordinator of the project on Medical Care and Medical Research of Bhopal gas victims sponsored by CHAI.)

Source: Sambhavna Documentation Center, Bhopal.

When money is not enough – Inadequate health care in Bhopal (2000)

Maya Shaw spent two months in Bhopal in the summer of 2000 investigating the health care provided in the Bhopal Memorial Hospital Trust community clinics. She was disappointed with what she found.

The Bhopal Memorial Hospital Trust (BMHT) was established in 1998, with money from the seizure of Union Carbide shares (Union Carbide assets in India were confiscated when they refused to appear in a criminal case) and a sizable contribution from the Indian government. Their financial resources are upwards of $87 million, with which they were directed to build a 260-bed hospital and 10 community clinics. As of today, it runs only five community clinics and the hospital has just become operational in the last two months. Even though gas survivors suffer chronic injuries that will last the rest of their lives, the Bhopal Memorial Hospital Trust is only obliged to provide free care at these sites for 8 years.

This report is focused on the care provided in the Bhopal Memorial Hospital Trust clinics. It looks at the clinic infrastructure, patient diagnosis and treatment, and doctors' attitudes and knowledge about gas-related health problems. Findings are qualitative, and are based on individual interviews with 14 doctors representing all five clinics, and the Director of the BMHT. It analyzes the ways in which the BMHT clinics fall short of providing lasting and comprehensive care. Ultimately, the report demonstrates that simply pouring money into construction of buildings, high-tech machinery, and medical staff is not enough, particularly when the services are only provided for eight years. The report argues that other interventions, such as community health education and environmental improvements, are likely to produce more lasting impacts.

The report concentrates on three topics:

I. Current health problems experienced by the survivors
II. Treatment provided in the Bhopal Memorial Hospital Trust clinics
III. Suggested interventions to improve health care in Bhopal [not included here – eds.]

I. The survivors' current health problems

It is estimated that currently 200,000 people in Bhopal suffer from debilitating chronic illnesses resulting from gas-exposure. The majority of these people suffer from respiratory illnesses such as fibrosis, bronchial asthma, Chronic Obstructive Airways Disease, emphysema and recurrent chest infections. In addition, because the gas-effected lungs are more susceptible to infection, pulmonary tuberculosis among the exposed population is significantly higher than the national average.

Other prominent health problems attributed to the gas are damage to the cornea of the eye, reproductive problems, post-traumatic stress syndrome and neurological problems. A plethora of studies on the health effects of the gas leak have documented these specific health problems. Studies on eye damage have found corneal ulcers, chronic conjunctivitis, deficiency of tear secretion, persistent corneal opacities, increased risk of eye infections, and chronic inflammation – resulting in persistent eye watering, photophobia (aversion to light), burning, itching, redness, and pain. Studies in reproductive problems have shown an increase in menstrual cycle disruption, leucorrhea (white discharge), dysmenorrhea (pain with menstruation), pregnancy loss and infant mortality. Studies on psychological effects of the disaster report high rates of anxiety, restlessness, grief, sleep disturbances and generalized weakness and fatigue. Studies on neurological damage have shown tingling, numbness, a sensation of pins and needles in the extremities and muscle aches.

II. The treatment provided in the Bhopal Memorial Hospital Trust clinics

Clinic Infrastructure

Each of the five BMHT clinics is located in a neighborhood near the factory that was seriously effected by the gas. At the time of the accident the majority of the affected population lived below poverty levels. Today, the majority of the survivors live in impoverished communities. To get to each of the clinics one must travel through narrow dirt roads flanked by worn-out and cramped houses. Long open tubes border the sides of the road in front of the homes, containing standing water and serving as depositories for human and kitchen waste. The BMHT clinics seem entirely out of place in these environments. Four of the five clinics are covered with glistening marble (the fifth is a dilapidated former Red Cross building) both inside and out. These brand-new opulent buildings represent a large expenditure of money.

Each clinic has a head doctor and two or three other staff doctor (three clinics have four doctors, two have three doctors.) Each clinic is staffed with an ophthalmologist and at least one general doctor. In addition, two of the clinics have Ear-Nose-Throat specialists, two have radiologists, and one has a pathologist. Each clinic is equipped with some means of laboratory diagnosis. The more sophisticated clinics have X-ray machines, machines for evaluating eyesight, spirometers and flow meters (to measure lung capacity).

Patient Diagnosis

Four of the clinics see 100 to 150 patients a day (one of the clinics sees only 50 patients a day). Many doctors reported that they typically see 50 patients during the 5½ hours that the clinic is open. On average, they spend less than five minutes with each patient. This is not enough time to conduct sufficient physical examinations. Physical exams are generally considered necessary to help doctors better understand patients' complaints, and also help them pick up on illnesses that will otherwise be missed until the patient is so ill they cannot be treated. Many doctors agree that a physical exam should include taking pulse and blood pressure, palpation of the abdomen, listening to the lungs and heart, and examination of the skin.

In a study carried out over five months in 1998, health workers from Sambhavna Trust (a nonprofit organization that provides health care and political advocacy for survivors) interviewed 474 patients seen in one BMHT clinic. Several patients reported that doctors at the clinic were reluctant to touch them, possibly because their bodies or clothing were viewed as dirty. In general, they found that examinations were minimal. 5.1% of patients had their pulse taken, 0.6% had their abdomens palpated, and 4.4.% had stethoscopic examinations. In my own interviews with BMHT clinic patients, seven patients told me that they received absolutely no physical examination at the clinics. They told me that to get physical checkups they must go to private doctors. Five women who complained of "white water discharge" said that they've never had a BMHT doctor ask them any questions about their symptoms; they were just given drugs.

Laboratory diagnosis is also minimal. The Sambhavna Trust health workers found that of the 474 patients seen in one BMHT clinic, only 6.1%

received blood tests, 1.3% received sputum tests (to test for TB), and 0.4% were given urine tests (to test for diabetes). In addition to being infrequent, lab tests are limited. Doctors at the BMHT clinics told me that in their lab tests they are looking for evidence of common diseases, such as TB, asthma, diabetes and anemia. Thus, they are only paying attention to health problems that they are familiar with and are likely to miss other long-term gas related problems that even the patients may not be aware of.

Treatment

In the summer of 1990, two nonprofit groups, The Bhopal Group for Information and Action (Bhopal) and Socially Active Medicos (Indore) released a document entitled "Evaluation of Some Aspects of Medical Treatment of Bhopal Gas Victims." In this report, they write:

"It is indeed shocking that while such a large population continues to be sick, a proper line of treatment of gas victims remains to be evolved. The prevailing symptomatic-supportive line of treatment currently being followed by doctors in Bhopal is indistinguishable from the line of treatment followed in the immediate aftermath of the disaster. As a result the gas victims can get only symptomatic temporary relief from their sufferings. While such a situation is understandable in the chaos, confusion and complete lack of information in the immediate aftermath, it is decidedly intolerable that after the passage of nearly six years, expenditure of crores of rupees and involvement of large number of professionals there is no effective improvement in the means for treating the gas victims."

Today, a decade after that statement was written, and after an even larger expenditure of money and investment of professionals' time, there has been no marked improvement in treatment provided to gas-exposed people. In a 1998 study of 502 BMHT health books, Dr. Rajiv Bhatia recognized that "Overall the drugs prescribed to the study population do not seem to be targeted to the organ system damage most likely to be consequent to the Union Carbide exposure as suggested by the available research findings...drugs are prescribed for short term symptomatic relief of nonspecific symptoms." Four BMHT doctors told me that the treatments they are providing to the gas victims are no different from the treatments they would provide to a non-gas-affected person with similar symptoms. This shows that the medical community has developed no therapies to address the injuries caused by the gas.

Lack of Treatment Protocols

The BMHT clinics do not have treatment protocols for specific gas-related illnesses. Thus, doctors' treatment is based on their own understanding of the survivors' health problems. This is problematic for two reasons. First, there are large discrepancies in the understanding of health problems; some doctors who have worked the community for a long time have fairly in-depth knowledge of the body systems effected by the gas, but other doctors know very little. Thus there will be discrepancies in care; the more knowledgeable doctors will provide therapies that they has seen are most effective, the ignorant doctors will be prescribing therapies that their colleagues have already seen to be useless or harmful. Second, combinations of certain medications can have a harmful effect. Therefore it is essential to have treatment protocols so that doctors know what drugs doctors in other specialties are prescribing. This knowledge of therapies would enable doctors to develop comprehensive treatment plans, incorporating the knowledge of each specialty.

Absence of Treatment Follow-up

The clinics do not have mechanisms for following-up on their patients. One doctor put it most succinctly when he told me: "It is up to the patients to return and let the doctors know how they are responding to treatment." This is not an appropriate way to treat chronic health problems. Because the majority of the survivors will be on medication for the rest of the their lives, it is essential that they are monitored closely to ensure that the medications are effective and do not cause secondary problems. In my interviews with patients I found that if the drugs they received from the BMHT clinics did not provide immediate relief from their symptoms they went to other doctors to get "better" drugs. In the treatment of chronic problems it is common that the effects of the medication will be seen only after continual use. Patients said BMHT doctors did not explain the importance of sticking with their treatment plans. Additionally, some of the medications prescribed can have harmful effects if combined with other medication. BMHT clinic doctors told me they are aware that patients see outside doctors for other drugs, but have no way of knowing what these drugs are. Comprehensive follow-up with patients would help doctors stay informed about patients' other medications and would help them avoid prescribing drugs that will interact negatively with these medications.

Prescription of Rational Drugs

Treatment at the BMHT clinics is focused on the distribution of pharmaceuticals. Some of the medications, such as corticosteroid injections, have significant potential toxicity, and have little or no demonstrated benefit. In the summer of 2000, Dr. Atanu Sarkar of the Catholic Health Association of India analyzed 101 prescriptions issued in one BMHT clinic. He compared the drugs prescribed with descriptions of patients' symptoms in their health books. He found that 26.3% of drugs given were harmful, 48.5% were useless to address the person's complaints, 7.6% were both harmful and useless. Only 17.6% of the drugs given were properly selected. In addition, he found that in a large number of cases (45.3%) when appropriate drugs *were* issued, the dosage was improper – people were instructed to take the wrong number of pills per day, were given a shorter course of drugs than needed, and were given less mg. per dose than needed. In some cases, he found that doctors were prescribing pediatric dosages for adults.

Doctors' Knowledge Regarding Health Effects of Gas

BMHT doctors told me that doctors were not chosen for the clinics because of any knowledge of gas-related problems, nor did doctors get any training regarding the specific health problems of the population they are serving. Therefore, it is not surprising that in my interviews doctors demonstrated little knowledge of the effect of the gas on people's health. One doctor summed up the lack of research and investigation: "We are totally in the dark. If such a disaster were to occur again we would not know what to do about it."

According to the BMHT Director, the clinics were developed to address the health problems identified in the Indian Council of Medical Research (ICMR) reports on health effects of the 1984 gas leak (the last study was terminated in 1990). Despite this, few doctors have basic knowledge of research that has been done on the health effects of the gas. In a fairly cursory search, I was able to find over 70 studies on the toxicological effects of MIC (Methyl Isocyanate was the main component of the gas released). Many of these were in prominent journals. These 70 include 26 studies of the human health effects of the Bhopal Gas Leak. Seven focus on ocular effects, seven on respiratory effects, four on reproductive effects, three on genetic effects, three on immune system functioning, one on psychological effects, and one on neurobehavioral effects. It appears that it would be rather easy for BMHT physicians to educate themselves about the special health problems of the population they are serving.

In an interview with the Director of the BMHT, I found him complacent about the lack of understanding among his staff about the health effects of the gas. I asked him why his staff did not know the ICMR findings. He answered that the doctors could have gotten the ICMR reports through their own efforts. He said that doctors are just treating problems as they come, and mentioned that many of the young doctors work in the clinics as a part time job (they also have private practice) and will probably leave the clinics in a couple of years.

BMHT clinics' doctors' lack of knowledge about the findings of research on long-term effects is evidenced by their disagreements about what health problems may have been caused by exposure to the gas. Regarding eye problems, one ophthalmologist told me that she did not know of any MIC related eye problems. She said most of the people she sees have allergic conditions (which she described as itching and redness), and loss of vision. She stated that there was not an unusually high incidence of eye problems among the gas-affected population. Another ophthalmologist said that watery eyes, itching and redness could possibly be a result of exposure, and that the gas caused corneal ulcers. A third ophthalmologist reported that the gas caused corneal opacities, burning, and corneal ulcers.

In regards to lung problems, one radiologist said that most patients' lungs are just fine and that there is no correlation between the increase in TB and exposure to the gas. A radiologist in a second clinic stated that the gas caused fibrosis, asthma, and bronchial asthma, and that the gas' damaging of the lungs made people more susceptible to TB. Seven other doctors mentioned specific lung damage caused by the gas, including bronchitis, emphysema, asthma, TB, fibrosis, and Obstructive Airway Disease.

Other health problems BMHT clinic doctors identified as possibly caused by the gas included: skin problems (2 doctors), cancer (1 doctor), ear infections (1 doctor), general weakness (1 doctor), psychological problems (2 doctors), gastrointestinal (2 doctors), and stunted growth of children (2 doctors).

If physicians do not have knowledge of the range of health problems that gas victims can experience, they do not know what problems to look for and cannot be effective diagnosticians. Doctors may misinterpret patients' symptoms because they are treating them like "normal" (non-gas-affected) people. Thus they lose the opportunity to catch a disease in its beginning stages.

Lack of understanding of gas-related health problems also allows doctors to down play patients' complaints. One doctor told me that there are not necessarily more health problems as a result of exposure to gas, it is just that there are more cases of illness diagnosed because there is now more free health care for this population. He said the major problem caused the gas is that people are psychologically disturbed and think that all of their health problems are caused by the gas. This doctor calls this problem "hypochondriosis". Three other doctors also said that most symptoms are only psychologically related to exposure…

Conclusion

One could argue that the critiques made of the BMHT in this report are overly judgmental and that suggestions for improvement are too ambitious. I would agree with these arguments if the $86 million came from any other source. The money that operates the BMHT clinics was seized from Union Carbide, and was earmarked to provide quality care to survivors. In effect, it is the survivors' money. It is fair for them to demand that it be invested in ways that provide them with the best possible rehabilitation for their health problems. Because the BMHT clinics and hospital only are obliged to provide free care for 8 years, it is not reasonable for them to use all of the money on fancy buildings and high-tech diagnostic machines that will be useless when they leave. I challenge the BMHT staff to use the remaining money in ways that will have a more lasting impact on the communities. If they teach people how to best care for their chronic health problems and improve community infrastructure to reduce spread of infection, the money will be well spent. If they continue to provide stopgap health care until all the money is gone, the BMHT will be just another injury to be suffered by the gas survivors.

Source: www.corpwatch.org

Five steps to recovery (1999)

Richard Mahapatra, writing for Down to Earth, *describes an example of ayurvedic treatment of a gas leak survivor at the Sambhavna Trust Clinic.*

***Panchakarma*, an ayurvedic treatment, holds the key to Bhopal gas victims' recovery**

Famida, 40, a resident of Bhopal's Jayaprakash Nagar was "anything but dead" for 14-odd years after the disastrous methyl isocyanate (MIC) gas leak in Bhopal in 1984. Two of her family members choked to death while running for their lives. She was left alive, but barely in a state to live a normal life. Blinding eye-burns and severe breathlessness, apart from the trauma of being forced to undergo an abortion, left her confined to her home.

Now, as she walks some 500 metres twice a week to an ayurvedic clinic administering treatment to the gas victims, she dreams of a life free of her illness. She was treated with panchakarma (literally, five steps), an ayurvedic treatment for flushing out toxic remains in her body, along with herbal oil massages. "It's a kind of second life they are giving me," she says, referring to Sambhavna Clinic – or the "clinic of possibility."

Set up by a band of dedicated social workers in 1995, the clinic's sole objective is to provide treatment to the 120,000 critically affected MIC gas victims.

It is a perennial battle against frequent abortions, anxiety, depression, insomnia, irritability, tuberculosis, respiratory problems, genetic deformation for generations, and impotency for those exposed to the toxic trail let loose by the MIC plant.

The death toll is tentatively put at 16,000, but more die every year, says a Sambhavna Trust report. Official counting of the disaster-related deaths was stopped in 1992.

"It was a hope against hope," says Satinath Sarangi, founder member of the Sambhavna Trust that runs the clinic. Till today, both the Union Carbide Corporation and the Union government have not revealed the type of gases, which leaked on that day. The former because they consider it a "trade secret" and the latter for unknown reasons. Hence, treatment to the gas victims is provided only on the basis of symptoms.

"We didn't know about modern treatment, and time was running out," says Madhusudan Despande, an ayurveda doctor at the clinic. The search for traditional treatment for chronic toxic exposure was not easy, as a need for such treatment had not risen earlier.

It took almost two years for a team to zero in on the 5,000-years-old *Sushrata Samhita, Bhaishaj Ratnavali, Rasendra sar Sangraha* and other standard *ayurveda* texts. To begin with, the clinic provided treatment for ailments like breathlessness.

According to the clinic's annual report, from March 1997 to March 1998, 2,273 victims had registered with the clinic, and visited 15,399 times. Some 40 to 50 gas victims visit the clinic every day. "The recovery encouraged people to come back to us," says Sarangi, "but there were relapses of diseases because we could not find a total cure."

Finally, in 1997, at a national seminar on ayurvedic treatment of gas victims, experts were consulted to make the treatment sustainable. *Panchakarma* treatment, a process through which the toxic contents are flushed out of the body through urine, stool and vomiting, was suggested and approved.

Along with *panchkarma*, nutrient herbs are injected into the body for absorption of toxic contents, while artificial diarrhoea is caused to clean the stomach. Similarly, medicinal herb water is injected through the nose to reduce panic attacks and depression, a common ailment among gas victims.

People visiting the clinic are free to choose between ayurvedic and allopathic treatment. However, as the illness is chronic among gas victims, both the treatments are clubbed together sometimes, though the ayurvedic treatment gets precedence. "It is very difficult to treat cardiovascular diseases. Ayurvedic treatment is preferred," says Dr Despande.

"After two years of treatment, I have started dreaming of a normal life," says Siddique, 40, who suffered from severe breathlessness and panic attacks frequently. "I am feeling well and can walk to the clinic easily without help. Earlier, even a slight exertion left me breathless. The number of panic attacks has been reduced, too," he says.

Thirty-year-old Sachin Ali and her son also get treated in Sambhavna Clinic. She was pregnant when the disaster struck. "I pumped in whatever resources I had on allopathic treatment but to no avail," she says. Both mother and son suffer from respiratory problems. "Within six month of treatment, I noticed a qualitative change in my health," she says.

To maintain quality and to cut costs, the herbal powder used in the treatment is produced in the clinic itself. Trained volunteers prepare 25 ayurvedic medicines with over 80 herbal ingredients gathered or purchased locally. Purchasing medicines externally is usually avoided. It is provided free of cost to the victims. Donations received by the clinic offsets costs. The clinic solicits individual donations through newspaper advertisements.

The three-year-old clinic seems to be attracting the attention of the gas victims. People working in the clinic say they have been getting new patients every day.

The clinic's volunteers visit the affected areas around the MIC plant daily and refer the victims for different types of treatment offered by the clinic. The clinic maintains a database on patients, and each patient is monitored individually to assess the recovery. "Soon we will publish a report on the effectiveness of ayurveda and yoga for treating gas victims," says Despande.

"It seems as if the people register at the clinic for a new life," says Rashida Bi, another victim who had been visiting the clinic for the last two years. After more than 14 years of illness, seeking a new lease of life is not asking for too much.

Source: Down to Earth, *published by Centre for Science and Environment, New Delhi, Vol. 8 No. 1, May 31, 1999.*

Healing Bhopal (2002)

Bhopal survivors, local activists, and a handful of medical professionals and social workers created the Sambhavna Clinic, which provides allopathic, ayurvedic and yogic treatment for the gas affected. Emily Polk describes the clinic

The Sambhavna clinic looks nothing like a hospital, but it saves lives.

It's been 18 years since a gas leak from a Union Carbide plant in Bhopal, India, killed 16,000 people. But the disaster continues to claim lives every day. According to official reports, half a million local people exposed to the toxic gases have poison still circulating in their bloodstreams. And many of them have come to rely upon a small, holistic clinic not far from the shuttered Union Carbide plant for help and hope.

The Sambhavna Clinic sees as many as 110 people a day suffering from internal organ damage, vision problems, shortness of breath, persistent coughs, menstrual irregularities, fatigue, depression, anxiety, and a host of other painful disorders. "Right from the beginning it was very obvious that modern medicine wasn't working for all the health problems [of the Bhopal disaster]," says Sathyu Sarangi, an engineer turned activist who helped found the clinic six years ago and currently serves as its managing trustee.

He launched the clinic with the belief that Indian officials, anxious to down play the disaster out of fear of "jeopardizing the investment climate," were not paying attention to side effects of the gas leak – tuberculosis, cancer, reproductive problems – showing up in survivors.

Sambhavna, which means "possibility" in Hindi, is housed in a two-story building that doesn't look like a hospital or a clinic. Potted flowering plants decorate the rooftop terraces. Fruit trees with benches underneath surround the grounds. Patients practise yoga on a terrace with walls decorated by the paintings of children. Inside, the clinic bears even less resemblance to conventional medical centers. The familiar odor of biocidal synthetic cleansers is absent; the clinic uses only plain water for cleaning. There are two rooms for ayurvedic massage and a library where information related to the gas accident is readily available.

In another room, staff members prepare more than 60 ayurvedic medicines, using herbal ingredients formulated according to this ancient Indian healing tradition. There are three cubicles for doctors, a computer room, a pathology laboratory, and a facility for regular cervical screening, Pap smears, and treatments for cervical cancer. Sambhavna is the only facility in the city to conduct regular Pap smears.

The clinic has survived mainly through appeals for donations in national and international newspapers. An international advisory group provides professional support to the clinic and serves as a link for fund-raising in other countries. Many of the staff also donate their services. After finishing her residency in community and family practice, Dr. Jaysi Chander spent six months volunteering at the clinic. "One of the most memorable and poignant moments for me," she says, "was when an elderly Muslim widow living in a slum in Bhopal offered a garland of flowers to me as a sign of her appreciation for the medical care I had offered her. She prayed for my good health and I for hers."

In addition to alternative therapies, the clinic practises an alternative form of organization. Instead of a traditional staff hierarchy, the clinic functions on a collective-management model. Each of the 25 staff members has equal input regarding decisions and equal responsibility for implementing them. The ratio of the maximum salary ($185 per month) to the minimum salary ($57) is barely more than three to one. Almost half of the staff are survivors of the accident.

Gary Cohen, co-coordinator of the U.S.-based Health Care Without Harm, a group that works to reform environmental practices in the health care industry, finds inspiration in Sambhavna. Cohen, who has worked with Union Carbide gas leak victims for years, says, "This clinic really represents the triumph of memory over forgetfulness. There's a profound way that both Union Carbide and the Indian government want to erase the memory of Bhopal, because it's an uncomfortable embodiment of the worst abuses of globalization. The clinic is a powerful symbol of people being empowered to defend and heal themselves."

The clinic conducts health surveys and continues to monitor deaths related to the gas exposure. Researchers use a questionnaire to conduct "verbal autopsies," interviewing a family member of the deceased to determine whether he or she died as a result of the Union Carbide leak.

"By all accounts, the poisoned night of December '84 is far from over in Bhopal," Sarangi says. "I am happy that, along with the survivors, we at Sambhavna refuse to go silently into this night. We have lit a lamp and continue to curse the darkness."

Source: Emily Polk, "Healing Bhopal," *Utne Reader*, November / December 2002, reprinted from *Whole Earth, Spring* 2002.

Unqualified medical practitioners: Private clinics in gas-affected Bhopal (2002)

This report by Jonathan Nash from November 2002 looks at the usage and usefulness of private, uncertified medical practitioners in Bhopal. Inexpensive and flexible, street doctors are often the only option for treatment of the gas affected.

Objectives of Study

The main objective of this study was to find out how private health clinics in gas-affected Bhopal operate on a day-to-day basis. The aim was to examine what general trends exist amongst the community of doctors who run these clinics, most importantly how patients are treated and how the doctors see health care in Bhopal with regards to the gas disaster. Having analysed the findings, the aim is to suggest possible solutions and areas worthy of further scrutiny.

How Research was Undertaken

Over a period of six days, eighteen health clinics were visited in the gas-affected areas of Chola, Berasia Rd, Qazi Camp and Tilazamalpur. These areas were chosen for their proximity to the most severely gas-affected communities, and also because of the unusually high concentration of health clinics in operation there. Doctors were interviewed and asked to answer a range of questions concerning their clinic, their patients and their treatment protocol. The duration of each interview was approximately one hour, and interviews were generally held in the manner of an informal conversation. I approached doctors as a student carrying out independent research. Whilst some doctors flatly refused to talk to me, the majority were willing, and after a few minutes of initial suspicion about my motives, began to answer my questions.

Limits of the Study

There are undoubtedly a number of problems with a study such as the one I have attempted to undertake. The fact that all findings are based on the answers doctors gave in a series of informal interviews has a number of drawbacks; the findings are the basis of opinion and as such must not be taken as hard factual evidence. It is hard to ascertain the degree to which doctors might have been exaggerating problems, or bending their answers to suit their own purposes. However, the purpose of this study is not to provide hard facts as such, but merely to facilitate a greater degree of understanding with regards to how private health clinics are run, and the repercussions that this is likely to have on the health of the gas-affected.

Findings

The findings of the study are displayed below and represent the answers given by the eighteen doctors who were interviewed.

1. Clinic age

The age of clinics ranged from two months to 36 years. In total, 40% of the clinics visited were open before the gas leak in 1984. Five clinics had been open for between 0–5 years, six had been open for between 5–20 years, and seven had been open for more than 20 years. Of these seven, three clinics had been open for 35 or more years.

2. Opening hours

Trends in clinic opening hours did much to reaffirm the image of the private health clinic as an easily accessible place for treatment – 14 clinics were open for seven days a week, and the rest for six. In terms of opening hours, the clinics fell into one of two categories. Generally, clinics were either open for a period of around 12 hours from 9 a.m. – 9 p.m., or they opened for a morning shift from 9 a.m. – 12 a.m. and then remained closed until the evening time, reopening between 6 p.m. – 10 p.m. What became apparent quite quickly however, was that opening times and the presence of a doctor generally didn't mean the same thing. In a number of clinics – for example Laxshdeep Muskan Hospital on Chola Road – the clinic stayed open all day, yet the doctor only came between 10 a.m. and 12 a.m. When asking if I could speak to the doctor, I was commonly told to "come back later" by people working inside the clinic. Whilst in some cases this was undoubtedly a tactic to avoid answering my questions, in the majority of clinics this genuinely was because the doctor wasn't present. This leads to the obvious question of exactly *who* these people are who run the clinic whilst the doctor isn't present. In two clinics I was told, "Come back later, the doctor isn't in" by people administering injections and prescribing medicines to patients. When I returned in the evening I spoke to the doctor, a completely different person. The concern here is that many of the clinics operate without the presence of a doctor for the majority of their opening hours, leaving the diagnosis and treatment of patients in the hands of unqualified assistants.

3. Number of patients seen per day

The number of patients seen per day ranged from 4–200 – the wide range presented is clearly a reflection on the varied opening times adopted by clinics. Figures of between 4–10 patients were given by clinics open for no longer than three hours per day. For clinics open both in the morning and evening and for those open all day, figures generally ranged from around

20–50 patients per day. No clinic estimated more than 60 per day, except for the Singhai Clinic on Chola Road, with an estimated 200 patients per day. This clinic was considerably bigger than most of the other clinics visited during the study, and had an adjacent room with roughly six beds on which patients could lie.

Assuming that around 40 clinics are in operation within gas affected areas (a conservative estimate), each receiving on average 30 patients per day, then we can calculate a figure of 1200 patients per day visiting private clinics. It is likely that a significant number of gas-affected patients will visit the clinics located near the centre of Bhopal surrounding Hamidia Road due to their proximity to places of work etc. Whilst this study only dealt with clinics in a number of gas-affected areas, the sheer proliferation of private clinics throughout Bhopal is undoubtedly an indicator that per day, clinics treat as many patients as the city's larger hospitals.

4. Location of patient base

Findings from the study do much to support the notion of private health clinics as places most frequented by the residents of nearby bustees and areas of poverty. All 18 doctors visited regarded the majority of their patients as residents of neighboring bustees, and almost all doctors commented on the problem of poverty when talking about their patients. Three doctors expanded their patient base to include villages outside of Bhopal, only one doctor claimed to have patients from all areas of Bhopal.

5. Patient symptoms

Doctors were asked to specify what trends existed in the age and sex of their patients, and how symptoms varied between age and sex. Responses were vague, with 16 doctors refuting the existence of any trends, claiming that "all types" visited their clinic, and that in general the same symptoms were evident throughout the range of age and sex. Of the remaining two clinics, the doctors regarded the majority of their patients as women, and generally dealt with gynecological problems. The general response to this question is an indicator that most doctors fail to separate patients into age and sex brackets with regards to both diagnosis and treatment – consequentially the specific needs of certain categories are disregarded in favour of a 'blanket' approach that covers all possibilities.

When asked to name the most common symptoms displayed by patients, doctors listed the following:

Common cold and coughs (13)
Fever (11)
Chest and respiratory problems including asthma and bronchitis (8)

Menstrual problems, most commonly menstrual irregularity (8)
Stomach problems such as loose motion and dysentery (5)
Malaria (5)
Eye problems such as conjunctivitis (3)
Liver problems (2)
Joint problems (2)
Ear problems (1)
Pneumonia (1)
Anxiety/Ghabarat (1)
Hypertension (1)

6. Patients' symptoms and the role of MIC

Doctors were then asked for their opinion regarding the role of MIC gas poisoning in the symptoms displayed by patients. Two doctors completely rejected the notion that the gas was a cause of health problems amongst the people of Bhopal. Whilst one of these doctors was an unprofessional, unqualified doctor who knew little about medicine in general, the other doctor, Anil Tilwani, was qualified, and ran one of the busiest clinics on Chola Road. He was quick to dismiss claims of widespread gas-related symptoms as "propaganda...," and stated forcefully that "there are no gas victims at all below the age of 50 years...." Tilwani's opinions worried me a great deal, and I found that many doctors followed a similar pattern though to varying degrees. It was common for doctors to regard gas related illness as a problem faced only by those of old age. Dr Prakash Issrani in Chola Road talked of "a new generation" of young patients, for whom gas related symptoms weren't an issue – he claimed that gas-affected patients made up a mere 50% of his older patients. The Shri Clinic in Tilazamalpur and the Farheen Clinic on Chola Road were two other places where I was told in no uncertain terms that no children were facing health problems with regards to the gas. It is undoubtedly the case that many doctors are uninformed about the continuing toxic legacy of MIC in Bhopal – the widespread belief that it is a "problem of the past generation" is a serious concern.

In general, however, there was a certain degree of agreement amongst doctors that the legacy of 1984 was still a prominent feature of health care in the city. In total, 10 doctors regarded the gas-affected as the majority of their patients; Dr N Singhai, situated on Chola Road, estimated that amongst his patients an overwhelming 90% were coming with gas-related health problems. In general, it was clinics opened prior to the disaster that put the most emphasis upon the problem of gas-related health issues, whereas the more recent clinics displayed varying attitudes toward the gravity of the effects upon the health of the MIC-exposed. Doctors with experience before

1984 generally commented on how trends in patients' health problems had changed after the gas leak; Dr Kishan Lal of the New Veenu Clinic on Chola Road had been working in the area since 1968, and claimed that respiratory problems have skyrocketed in the 18 years since the disaster. For many doctors, it was problems of endemic poverty rather than gas-related illness that formed the basis of their patient's health problems; Dr Tilwani was adamant that Chola's health care network was dealing with universal issues of poverty such as lack of sanitation and unclean water, and that patients with identical symptoms would exist in neighborhoods like this throughout the world. Even for doctors who did emphasize the issue of gas-related illness, they were generally keen to stress the crucial role of problems resulting from impoverished living conditions.

7. Perceived effects of MIC exposure

Doctors were asked their opinion on how exposure to the MIC gas had affected people's health – what were the main symptoms caused by MIC exposure, and how it affected the different organs in the body. Only 4 doctors seemed familiar with the term "MIC gas," and were able to talk about the effects in accurate comprehensive terms. Each of these four doctors explained how MIC had affected every part of the body by finding its way into the patients' blood stream and tissues. Two of these doctors, one being Dr K. P. Soni of the Care and Cure Clinic in Berasia Road, talked of a reduced resistance to disease that combined with poverty to cause a major health crisis amongst the gas-affected communities. Generally however, the majority of the remaining doctors failed to describe the effects of MIC exposure in accurate terms, and were only able to list a number of perceived gas related symptoms, without any form of medical explanation.

When asked to explain the symptoms of MIC exposure, the following were mentioned:

Respiratory, lung and chest problems (12)
Eye problems (9)
Menstrual problems (5)
Anxiety/Ghabarat (3)
Stomach problems including dysentery and loose motion (2)
Hypertension (2)
Fever (2)
Liver problems (1)
Joint problems (1)
Fatigue/weakness (1)
Cardiovascular problems (1)
Kidney problems (1)

It is clear from these answers that few doctors were able to provide a comprehensive list of the symptoms of MIC exposure. Most focused solely on respiratory problems, and even in this area were vague in their description of the exact type of problems encountered. The general lack of knowledge concerning the full effects of MIC exposure is a grave concern; it will only serve to further prevent the gas-affected population of Bhopal from receiving adequate treatment for the wide range of long-term health problems they face.

8. Treatment protocol

After discussing their patients and the range of health problems within local communities, conversation turned towards the kind of treatment administered in the clinic. Doctors were asked first of all what type of medicines are most frequently administered in the clinic. Allopathy was undoubtedly the favored method of treatment in the clinics; nine doctors used solely allopathic medicines, and a further four favored allopathic but occasionally used ayurvedic medicines. Three doctors used homeopathic treatment and two used solely ayurvedic. A number of reasons were given for the widespread use of allopathy – Dr V. K. Anand of the Anand Clinic on Chola road claimed that allopathic drugs such as antibiotics were commonly used because they gave the quickest results. G. S Hameshal clinic on Berasia Road echoed this sentiment when its doctor informed me of his opinion that many homeopathic doctors were switching to allopathy for its qualities of "instant relief." The sheer number of dispensaries dealing in allopathic drugs in and around the areas visited is undoubtedly another major reason for the tendency to favour allopathic treatment.

- **Length of prescription**
  When asked about the length of the average course of treatment administered to patients, figures ranged from one day to one week. Eight doctors prescribed just one day of medicine to patients, the remaining doctors prescribed between 2–3 days, and two doctors prescribed 4–5 days. The reasoning behind the tendency to prescribe such short courses of treatment was commonly attributed to the patients' inability to afford anything longer than a few days of medicine at one time. The trend towards short dosages of allopathic drugs such as antibiotics was therefore striking during the course of this study, and it is a concern since the capacity of such a treatment protocol in dealing with long-term health problems is limited at best.

- **Cost of consultation**
  Clinics generally fell into one of two categories – those that prescribe medicines, and those that dispense medicines. Dispensing

clinics were often run as glorified chemists – after a brief examination by stethoscope, a handful of drugs were passed over and for a sum of roughly 10 rupees the patient left with a one day course. The trend seemed to be that in non-dispensing chemists, the doctor would charge a figure of around 20 rupees, purely as a consultation fee. Such was the case in Dr Sheli's clinic in Qazi Camp. She complained that unqualified "quacks" were taking the business away from qualified doctors like herself, purely because they were willing to consult and dispense for around 10 rupees. A commonly voiced concern amongst qualified doctors who didn't dispense medicines themselves was that the dispensing clinics often use low quality or fake medicines to cut their costs. This point was brought up by Dr V. K. Anand on Chola Road amongst others. The implications of such an activity are clear, and it is an area that demands closer research.

– **Commonly Prescribed Drugs**
  Having asked a few general questions about treatment protocol, I attempted to find out exactly what drugs were most popularly prescribed in each clinic. The response to such probing varied greatly – a number of doctors were extremely reluctant to divulge such information, insisting that treatment was "symptomatic" and refusing to say more. Others were however more cooperative, and the following results were collected by asking doctors how they generally dealt with a variety of symptoms.

General Pain:

The most commonly administered painkillers were aspirin, paracetamol and ibruprofen, often prescribed in combination. A considerable number of doctors also chose to administer antibiotics for general pain relief, commonly choosing ciprofloxacin and doxycycline. Two doctors mentioned the use of valium for common pain relief.

Chest and Respiratory problems:

Antibiotics, cough syrups and bronchodilators were the most common form of drugs administered for chest and respiratory problems. The most frequently named antibiotics were amoxicillin and cefardroxil. Cefardroxil is however primarily used to treat urinary tract and throat infections; the recommended course is at least one week, yet as was mentioned earlier, antibiotics are generally administered for just a single day. The majority of clinics had a large range of cough syrups, most commonly benadryl expectorant. Syrups often looked cheap and rarely displayed a list of ingredients on their packaging. Doctors who mentioned other drugs named corticosteroids, dexamethasone and chloropheniramine, used for the treatment of allergies and asthma.

Fever:

For the treatment of fever all doctors who divulged information seemed to use a combination of painkillers such as ibruprofen and paracetamol with antibiotics – all doctors were using ciprofloxacin as the antibiotic of choice. It was common for doctors to administer up to three different types of painkillers, for example, dispirin, paracetamol and ibruprofen, together. This mixing of medicines, administered to patients without any form of advice, is a grave concern.

Menstrual Problems:

Doctors favored a wide variety of tonics and syrups for the treatment of menstrual problems. Iron and calcium tonics were popular – it was common for allopathic doctors to treat menstrual problems with ayurvedic medicines. Ashoka syrups, produced in India by BVP and PRL were the most commonly stocked drugs for the treatment of menstrual problems. Femiplex ayurvedic tablets were also used in a number of clinics. As with the treatment of chest and respiratory problems, the beneficial qualities of many of the syrups looked questionable, and again a large majority failed to display an ingredients list.

Eye Problems:

Eye drops were used by all doctors who talked about the treatment of eye problems – pyrimon and betanasol were the most commonly prescribed brands. One doctor claimed to use ciprofloxacin to treat eye infections.

Liver Problems:

Few doctors chose to comment on their treatment of liver problems. The majority of those who did mentioned the ayurvedic syrup Liv 52 as their principal form of treatment. Other ayurvedic syrups mentioned were G-Liv and Pro-Liv.

General Comments:

The doctors' responses to my questioning surrounding their most commonly prescribed drugs bring up a number of important issues. The widespread mixing of numerous different types of drugs, in particular antibiotics and painkillers, is a cause for concern. In one clinic I visited, the doctor was using a mortar and pestle to crush 4 different types of painkiller into a chalk, which he proceeded to give to a 15 month old child. Combining with this issue is the problem already raised that patients are rarely given the recommended full course of medicines, which is in many cases likely to prevent effective treatment. In general, their responses seem to further clarify private clinics as wholly ineffective in dealing with the complex long-term health demands of the gas-affected.

- **Follow-up treatment**

  Doctors were asked how they deal with patients who return after the treatment has failed to bring relief. Here, the general response again uncovered deep inadequacies within the treatment protocol – only three doctors showed any sign of an organized system of follow-up treatment. Dr Sheli in Qazi Camp stressed the importance of regular checkups for a period of one month for all gynaecological problems, whilst Dr Kishan Lal of the New Veenu Clinic on Chola Road made all returning patients take a series of radiology and pathology tests. The majority of doctors failed to offer effective follow-up treatment, generally opting for a repeat one-day prescription of antibiotics. It seemed common practice to administer an antibiotic injection to patients returning with problems that a course of tablets had not managed to fix. The widespread failure to keep any form of records concerning patients' medical history is the major stumbling block for administering effective follow-up treatment – astonishingly, only three doctors kept adequate medical records cataloguing the date, patient name, symptoms and treatment administered. One of these doctors had allotted each patient a "serial number," but such a system was not in operation anywhere else. Often the only record of treatment was a scrawled note for the patient to hand over to a dispensing chemist, quickly lost and forgotten. This widespread failure to accurately catalogue medical records has major consequences for health care in gas-affected Bhopal. The long-term nature of many of the gas-affected's health problems calls for a close monitor of past treatment – without such careful consideration, the scope for inappropriate treatment is vast.

- **Referrals**

  If symptoms remained after two consecutive visits, it was common for doctors to refer their patients to a hospital for further treatment. Doctors were generally vague about which hospitals they referred patients to, but the most common choice seemed to be Hamidia Hospital, as well as private nursing homes. With regards to referrals, inadequate patient records are again a major cause for concern – it seems to be the case that the clinics have no form of communication with the larger hospitals through which they can be informed of a patients' medical history. Patients were generally given a phone number by the clinic and told to leave – this cold, impersonal treatment was a feature of almost all the clinics I visited. Without adequate communication between clinics and larger health care centres such as nursing homes and hospitals, referral by clinics is only likely to further delay patient's access to effective treatment.

9. Medical Representatives

Doctors were asked a number of questions concerning medical representatives from drug companies. All doctors commented on the reasonably regular visits of medical representatives – figures ranged from one per month, to five per week, but were generally around one per week. A wide variety of drug companies were named, but a few seemed to dominate. Himalaya, an Indian company specializing in ayurvedic medicine, and Cipla, an Indian-based manufacturer of allopathic medicines, were the most commonly mentioned companies. Other companies were Glaxo, Ranbaxy, Olympic and Bayer. A number of doctors such as Dr V. K. Sahu of the Vikash Clinic, Chola Road, bought all their medicines from local dispensing chemists rather than visiting representatives. However, the general trend is undoubtedly that medical representatives play an active role in the day-to-day dealings of private health clinics in Bhopal.

Suggested Areas for Improvement

Health care in private health clinics in Bhopal is inadequate in a number of ways:

- Doctors are often unqualified and/or operating as a business rather than a place of treatment;
- Doctors fail to keep adequate patient records and have no idea of a patients' medical history;
- Treatment generally revolves around short-term doses of allopathic medicine;
- Medicines are often mixed together and given without adequate information;
- Doctors do not have an adequate understanding of how MIC has affected gas victims;
- Doctors are unaware of many of the long-term health problems related to MIC exposure.

These are factors that combine to form a major stumbling block towards effective treatment. In order to improve the health care situation in private health clinics,

- Doctors must be educated on the effects of MIC to realize that an ongoing crisis exists;
- Pressure must be placed on doctors to begin adequate record keeping;
- A system of greater communication must be established between clinics and hospitals;
- Pressure must be put to weed out unqualified doctors;
- Patients must be educated about health care and medicines and encouraged to keep records themselves.

Source: By Jonathan Nash, November 2002.
From the Sambhavna Documentation Center, Bhopal.

PART 3: CHANGES

10

Movements

Bhopal activism, 20 years strong

"BHOPAL WILL NOW come alive in every corner of the world" declared a 1985 protest song as the anger and injustice that has come to characterize Bhopal was beginning to crystallize. Many years, songs and protests later, as delayed justice has compounded the injury of the disaster, it is *still* alive. Survivors have had to fight tenatious legal and/or extralegal battles for every single crumb of medical and economic rehabilitation that they have received, and a trial to find the truth about the disaster – along with the punishment of the guilty – is still pending. Organization and agitation have been a constant imperative and have drawn on many disparate dialogues and movements. Issues of health, human rights, environmentalism, international law, corporate accountability, women's rights, poverty and discrimination are all relevant. Other communities recognize their problems in the history of Bhopal, as the impunity of corporate actors becomes an increasingly familiar story.

Unlike most cities in India, Bhopal had very little history of grassroots militancy prior to 1984. The collaboration of the Nabab and Begum rulers with the British effectively kept the citizens of Bhopal away from the revolutionary fervor of the 1857 uprising, and thereafter well into the twentieth century from other movements as well. The first demonstrations after the gas disaster were spontaneous – the structures and techniques of protest developed slowly in the aftermath. In their struggle against one of the world's most powerful corporations, the people of Bhopal – who owned little and had lost much – truly began with nothing.

At heart, beyond the many different kinds of violations that characterize the disaster, Bhopal is a story of the powerless fighting back against the powerful. Supported over the years by a ragtag bunch of activists, and occasionally by international organizations (i.e. the Red Cross in the immediate aftermath, Greenpeace in the past years – see their report in the chapter, "Contamination," and their Bhopal Principles in the final chapter of this book), and most recently, Amnesty International), every tiny victory has been the result of enormous amounts of work on the ground. The gas disaster also forced many in the affected communities, particularly women, to learn about and face the dynamics in the outside world that brought Union Carbide to their doorsteps. For some of the Muslim women who fled the gas, the evening of 3 December 1984 was the first time they left their homes without a veil; many have since discarded the burkha permanently. With the disruption of family structures during the gas, many Hindu and Muslim women were forced to look for work outside the home, and through necessity have stayed outside it to agitate for justice. Although men took the leadership roles in the first few years, women – politicized more slowly and usually through dire necessity – quickly became, and have remained, the backbone of activism in Bhopal (as described in the following excerpts).

Currently there are five survivors' organizations in Bhopal, with active memberships of approximately 10,000 people. Although the length and frustration of the struggle has sometimes caused discord between the different groups, the grassroots organization and politicization of these womens' unions and other groups remain unique and powerful. Unsupported by the local organized political parties or government officials, their weekly meetings, marches to New Delhi and networking with similar communities worldwide continue to chart new territory.

Similarly, the expansion of the activism about Bhopal onto the internet, in newspapers, and into the barely tested waters of shareholder activism, have caused a recent explosion of interest and action, much of which coincided with the twentieth anniversary year. Recently resolutions on Bhopal were introduced in the United States

Congress, two survivors and leaders of one women's union won the prestigious Goldman Prize for the Environment (as related by Penny Wark below, and in their own words in the chapter on "Surviving"). As the summary by the student activism coordinator at the end of this chapter illustrates, agitation for Bhopal is today reaching new heights internationally.

The whole world trembles (1985)

People's movements in Bhopal have been the relentless motivator for justice that has carried the issue for two decades. Songs like the one translated here, adapted from Balli Singh Cheema, have given strength and voice to these protests.

With flaming torches we march / with flaming torches in our hands / we march / we the people of Bhopal / now we will conquer darkness, / we the people of Bhopal / the huts ask and the farms ask us too / how long we will be oppressed, / we the people of Bhopal / every obstacle screams as we kick it out of our way / we make music with our shackles, / we the people of Bhopal / see friends, the morning looking pale in these times / we will fill it with the colour red, / we the people of Bhopal / knowing that you cannot get anything without struggle / we are fighting a battle, / we the people of Bhopal / we have tied our shrouds on our heads / and we hold swords in our hands / we have come out to look for our enemies, / we the people of Bhopal

Bhopal will now come alive in every corner of the world /and we will continue with our battle, /we the people of Bhopal.

Source: Sambhavna Documentation Center, translated and adapted from a song written by Mr. Balli Singh Cheema.

Bhopal lives: The lifting of the veils (1996)

Suketu Mehta, in another part of his series, examines the women's movements in Bhopal in the years after the disaster.

In the years after the poison cloud came down from the factory, the veils covering the faces of the Muslim women of Bhopal started coming off too.

The Bhopal Gas Peedit Mahila Udyog Sangathan (the Bhopal Gas-Affected Women Workers' Organization), or BGPMUS, is the most remarkable and, after all these years, the most sustained movement to have sprung up in response to the disaster. The BGPMUS grew out of a group of sewing centers formed after the event to give poor women affected by the gas a means of livelihood. As they came together into the organization, the women participated in hundred of demonstrations, hired attorneys to fight the case against Carbide as well as the Indian government, and linked up with activist movements all over India and the world.

On any Saturday in Bhopal, you can go to the park opposite Lady Hospital and sit among an audience of several hundred women and watch all your stereotypes about traditional Indian women get shattered. I listened as a grandmother in her sixties got up and hurled abuse at the government with a vigor that Newt Gingrich would envy. She was followed by a woman in a plain sari who spoke for an hour about the role of multinationals in the third world, the wasteful expenditure of the government on sports stadiums, and the rampant corruption to be found everywhere in the country.

As the women of Bhopal got politicized after the gas, they became aware of other inequities in their lives too. Slowly, the Muslim women of the BGPMUS started coming out of the veil. They explained this to others and themselves by saying: look, we have to travel so much, give speeches, and this burkha, this long black curtain, is hot and makes our health worse.

But this was not a sudden process; great care was paid to social sensitivities. When Amida Bi wanted to give up her burkha, she asked her husband. "My husband took permission from his older brother and my parents." Assent having been given all around, Amida Bi now goes all over the country without her veil, secure in the full support of her extended family.

Her daughters, however, are another matter. Having been married out to other families, they still wear the burkha. But Amida Bi refuses to allow her own two daughters-in-law, over whom she has authority, to wear the veil at all. "I don't think the burkha is bad," she says. "But you can do a lot of shameful things while wearing a burkha."

Half of the Muslim women still attending the rallies have folded up their burkhas for ever.

Source: *Village Voice*, December 3 and December 10, 1996.

Resistance on the ground: Survivor organizations fight for their rights (1994)

Amrita Basu looks at the emergence of women's movements in Bhopal, and at the mothers who have mobilized to stage large-scale protests and actions for their cause.

In February 1989 the Indian Supreme Court announced that it was proposing a final settlement in which the government would accept from UCC $470 million on behalf of the victims and discharge it from all civil and criminal liability for the disaster... Within a week of the Supreme Court decision the BGPMUS mobilized 3,500 people to travel to Delhi where they ransacked the Union Carbide office. Press reports highlighted women's militancy: one woman reportedly snatched a gun from an armed guard, while another chased a policeman until he jumped onto a passing bus for safety.[4]

Over the next several months the BGPMUS continued to organize large-scale protest. One of the most ingenious was a "Quit India" demonstration in August 1989, on the forty-seventh anniversary of the 1942 Quit India movement demanding immediate independence from the British. Three thousand BGPMUS members demonstrated outside the UCC plant in opposition to the court decision and demanded that UCC terminate all activities in India. Newspapers reported police violence resulting in serious injuries to over 300 women; over a thousand women courted arrest.[5] The demonstration elicited widespread support; a predominantly Muslim organization had exposed the way the Congress Party had surrendered to foreign interests while claiming concern for the most vulnerable and marginal groups.

The BGPMUS was responsible for many of the gains credited to the Janata Dal, the coalition that became the national government after winning the election in 1989. Since its criticisms of the Supreme Court settlement came just before the 1989 parliamentary elections, the Janata Dal was receptive to grievances against Congress. In November 1990 the newly elected government initiated a rehearing of the settlement based on the review the BGPMUS had filed. No government had addressed the question of how gas victims would survive in the many years before they obtained a final settlement. In June 1988 the BGPMUS filed a petition in court demanding interim relief for the gas victims. In March 1990 the Supreme Court ordered the MP government to pay monthly sums of 200 rupees ($12) per month for three years to the 500,000 people it determined were in the vicinity of the gas disaster on the night of 2–3 December 1984. But in March 1990, the Hindu nationalist BJP was elected to head the MP state government. A BJP government would administer relief and rehabilitation and determine policies for Bhopal's Muslim population.

The BJP's stance on the UCC tragedy provided it with additional ammunition against the Congress Party during the election campaign; its populist stance on relief for the gas victims helped broaden its appeal to include disempowered groups. After coming to power, however, the situation changed. There were more claimants on state resources, which in turn complicated the BJP's decisions about how to use resources designed for relief. Moreover, the BJP's traditional constituency of Hindu shopkeepers and traders had nothing to gain from an activist stance with respect to the gas victims.

Contrary to its promises, the BJP government neglected to involve voluntary organizations in the distribution of interim relief. Instead, it packed an advisory committee on interim relief with members of Hindu communal groups.

Furthermore, the state government's distribution of interim relief was slow, inefficient, and dishonest. Contrary to the government's promise that all eligible people would receive interim relief within a month, three months later only 42,000 claimants – less than 10 percent of those who were eligible – were receiving interim relief. A year later, then chief minister Sunderlal Patwa admitted that about 100,000 eligible victims were still not receiving interim relief.[8] According to Khan, one reason for this delay was that public hospitals had not even examined hundreds of thousands of claimants.

And as government officials admitted, about half the people to whom they sent notices never received them. Yet officials refused to disburse funds to these claimants.

Numerous gas survivors complained of extensive corruption in the distribution of interim relief. Rabiabia, a resident of a slum called JP Nagar, said that middlemen charged 1,000-1,400 rupees ($60-$84) for papers demonstrating her eligibility; the men who delivered notices to their homes also demanded bribes. Atiyabia, a BGPMUS activist, complained that the government refused to divulge the names of beneficiaries although this would have checked corruption. Many people attributed the government's secrecy to its use of relief designed for Muslims as a form of patronage of Hindus.

Recipients of interim relief also complained of difficulties in collecting their monthly payments. They had to collect their checks from distant places and spend hours waiting in line. One person was not authorized to collect funds on behalf of the family, so every month each family member would spend a day collecting a small payment. The government refused the

BGPMUS's request to either deposit checks directly in the recipients' bank accounts or deliver the checks to their homes.

Given the way bureaucracies function, the corruption and delays that characterized the distribution of interim relief might be considered unavoidable. However, while the government effectively controlled the bureaucracy in other situations, it seemed indifferent to the fate of poor Muslims. Conversely, the lower ranking Hindu government officials who benefited from the politicization of the relief distribution process were important elements of the BJP's constituency.

The BGPMUS attracted several thousand people when it staged demonstrations in 1990 and 1991 to protest the faulty distribution of relief and the inadequacy of medical care for the victims. However, activists became discouraged by several developments. The new Supreme Court judgment announced on 3 October 1991 largely upheld the 1989 settlement. It declared that if the settlement fell short of the amount to be paid to the victims, the Indian government – rather than UCC – would pay the deficit. Although it removed UCC's immunity from criminal prosecution, the judgment once again denied victims the right to be heard. Rather than providing new guidelines for evaluating the extent of injury and death, it relied upon the seriously flawed estimates ordered by the state government years earlier.

The MP government arrived at ludicrously low assessments of the number of gas-related injuries. Shortly before the 1989 elections, when Congress realized that the Supreme Court decision would be reviewed, it ordered the MP government to assess the medical conditions of gas survivors. The doctors' conclusions were designed to exonerate the Congress government and confirm the Supreme Court's decision. The MP government's medical report stated that only 19 people had been permanently and totally disabled; 155,000 people whose records had been examined had suffered no injury at all.

Consider these results in light of the evidence: 40 tons of deadly MIC were sprayed on the city of Bhopal. Over 125,000 people lived within a 3 kilometer radius of the city; a dose of 0.02 ppm of MIC is considered lethal. Although there was public outcry when the state government issued the report and many doctors contested the findings, the Supreme Court cited them in its 1991 judgment. UCC later justified the low settlement by referring to the results of the government survey.[9] The BJP government defended these findings.

Footnotes

4. Claude Alvares, "Bhopal's Fighting Mothers," *Patriot Magazine*, 20 August 1989.
5. Ibid.
8. *Sunday Observer*, 24 August 1991.
9. When the results of its survey were greeted with derision and scorn, the government admitted that its survey methods had been flawed and agreed to reexamine its results. However, it did only a random check of 10 percent of the cases, and the figures it resubmitted to the court would have a bearing on only 5,000 cases. Furthermore, in rehearing the case, the Supreme Court referred back to the findings of the unrevised study.

Source: Amrita Basu, "Bhopal Revisited: The View from Below," *Bulletin of Concerned Asian Scholars*, Vol. 26, Nos. 1-2, January – June 1994.

People's movements in Bhopal (1994)

Satinath Sarangi, an activist in Bhopal since 1984, presented a history of agitation in Bhopal in testimony before a session of the Permanent People's Tribunal on Environmental and Industrial Hazards.

Spontaneous Protests

The collective response of the survivors over the last ten years appears to have gone through three distinct phases – spontaneous protests in the immediate aftermath, organized under middle class leaders for the following two years and finally the formation of survivor led organizations. The first started with an angry march and a gathering of over one thousand survivors at the factory gates of the Union Carbide on the morning of the disaster. They had hardly any information on who was running the factory and how, and even less on its hazardous nature. When the marchers reached the factory a decision was taken to burn it down. The factory officials in panic spread a rumor that the gases had started leaking again. In confusion, the crowd fled away from the factory in a re-enactment of the previous night's mass panic. Spontaneous collective protests, mostly leaderless, continued in different communities without any overall inter-community organisation. In small groups, survivors demonstrated at government offices calling for medical care, monetary assistance and immediate hanging of the killers of Bhopal whose names, and particularly that of the chairman of the corporation, were widely known within the first three days.

With the declaration of Operation Faith – the neutralization of the remaining toxic chemicals – there was once again widespread fear and chaos from 10 December 1984. Over 400,000 survivors left Bhopal in less than two days. Small numbers opted to stay at the government relief camps set up in new Bhopal. The camps provided a place for the survivors from different communities to come together and protest against the near absence of government efforts to provide relief and care. Hundreds of people marched to the governor's residence on 16 December and then again two days later. A small number of Bhopal activists played a role in facilitating collective decisions and action. By the end of the month these activists were to arrogate more powers to themselves and initiate the second phase of the movement in Bhopal in which spontaneous protests found encouragement only on rare occasions.

Organized Response

Within the first week of the disaster, about thirty individuals with varying leftist persuasions met in two groups to found two organizations – Nagarik Rahat Aur Punarvas Committee (NRPC) and the Zahreeli Gas Kand Sangharsh Morcha (ZGKSM or Morcha for short) with distinct agendas. Though a few locals were involved in the founding of the organizations, out of town and new-Bhopal activists played a dominant role in outlining the respective "politics" of the organizations. For NPRC, provision of relief and rehabilitation was to be the main issue. The Morcha, or more correctly its leaders, stressed the need for a political organisation of the survivors that would take up issues of justice, access to scientific information, medical care, and legal intervention. Though there was reason and space for both organizations to coexist and support each other's work, internecine conflicts started brewing from their inception.

Survivor-led Organizations

Left to fend for themselves after the exodus of the middle class leaders and activists, the survivors were soon to organize themselves to continue with the struggle in Bhopal. Initially concerned with immediate problems of jobs, pensions for destitutes and regularizing of employment at rehabilitation centres, the four organizations that formed were soon to take up medical care, monetary relief, criminal liability, compensation, environmental rehabilitation and corruption by government officials as their rallying points. The organizations grew in strength and effectiveness and two of them – Bhopal Gas Peedit Mahila Udyog Sangathan and Nirashrit Pension Bhogi Sangharsh Morcha – had over 100,000 members within the first two years of their formation.

While these organizations resembled the earlier ones in size, range of concerns and ability to pressure the government, as things turned out, their resemblance did not end there. In fact, these organizations resembled traditional Indian extended families in many of their features and their leaders had even less respect for democratic functioning. Conflicts between organizations grew as their leaders competed with each other to be the sole representative of the survivors. In contrast with organizations formed earlier, women outnumbered men several times over. Defying traditions of their respective religions and family bondage, Muslim and Hindu women survivors played an active and sustained role in the organizations. However, the two largest organizations came to be dominated by men with few scruples about usurping female power for their personal glory.

Conscious of the need for research, documentation and monitoring activities as they were, the survivors' organizations in the third phase did not have the necessary skill and training to be able to carry them out. Dissemination of information was limited to the minimum required for immediate mobilization of people around particular issues. The involvement of a large number of women presented the possibility of organizing their wide ranging production skills into income generating cooperatives. However, the organizations chose to depend upon and pressure the government into providing jobs to the Bhopal survivors. Even the closure of the sewing centres by the government in July 1992, did not prompt any initiatives for helping survivors to become self supporting. With time the two major organizations became involved with party politics and the popular response to the disaster was almost back to the traditional politics of Bhopal.

Given these serious shortcomings, the latter day survivors' organizations have had significant achievements. Much needed monetary assistance from the government, modification of the infamous settlement order, withdrawal of criminal immunity from Carbide and its officials and most government relief and rehabilitative measures have been made possible through their legal and extralegal interventions. Above all, through their continuing marches and rallies demanding justice and a better deal for the survivors, they have kept Bhopal alive in the public mind.

Source: from *Righting Corporate Wrongs*, evidence presented to the Permanent People's Tribunal, London Session, Autumn 1994.

Should corporations get away with murder? (1995)

As the second post-disaster decade began, increasing effort to inform a wider audience took form, as exemplified by this full page ad that appeared in the New York Times. *Endorsed by 25 environmental and human rights organizations, it tries to expose Carbide's duplicity in dealing with the Bhopal issue.*

Thousands of men, women and children were killed when Union Carbide gassed Bhopal, India. A decade later, tens of thousands more are still crippled and dying. Why is this corporation still in business?

Union Carbide, a $5 billion chemical company chartered in New York state, settled out of court for less than $800 per claimant when its pesticide plant in Bhopal, India, released a lethal cloud on that city of 800,000 in 1984.

Wall Street's appreciative response was a $2 jump in Union Carbide shares.

But justice still eludes Bhopal.

Health surveys reveal that a decade after the leak 204,000 people still suffer from gas poisoning. The official death toll has climbed to over 6,000. Criminal charges in India still stand against Union Carbide and two Asian subsidiaries. And an arrest warrant has been issued for Union Carbide's chairman and CEO at the time of the disaster, Warren Anderson.

The charge? Culpable homicide and grievous harm.

Will Americans press for justice in the world's worst industrial catastrophe? Or will we let Union Carbide teach other U.S. corporate bosses how to get away with crimes against communities overseas – and here at home?

This isn't the first time in the twentieth century that a corporation has been accused of criminal acts.

After WWII, German arms-maker Krupp was forced to divest, and chemical giant I. G. Farben broken up, because they were implicated in slave labor and other war crimes.

Union Carbide knew, from past accidents at the Bhopal plant and its own internal safety audits, of serious flaws in the design and operation of this extremely hazardous plant. Yet it did virtually nothing to avert disaster.

Most state corporation codes, including New York's, provide for dissolving a corporation that causes mass harm.

So why is Union Carbide, perpetrator of the worst industrial disaster in history, still allowed to pursue business as usual?

How Union Carbide Got Away With Slaughter In Bhopal.

American corporations have become increasingly adept at using bankruptcy laws to protect their profits from those they maim or kill. But Union Carbide, shielded by a huge chemical industry PR campaign, needed no such protection.

Instead, Union Carbide sold off its consumer product lines in 1985 and 1986 before boycotts could erode their market share. And it paid out a big hunk of its assets to Carbide shareholders, thereby placing those assets beyond the reach of thousands of victims in Bhopal.

After "down-sizing" its workforce by 90%, Union Carbide rang up 1994 profits of $379 million.

Meanwhile, both the corporation and its Chairman at the time have refused to appear in court in India to answer criminal charges. And the U.S. and Indian governments appear to have shelved any effort to make them do so.

Tens of Thousands In Bhopal Continue To Suffer From Union Carbide's Reckless Acts.

More than 200,000 residents continue to suffer from their toxic exposure, many so terribly that they are unable to support their families. Ten to fifteen more people die each month from medical causes related to the catastrophe.

Local organizations have repeatedly asked for information about Union Carbide's toxins that might help relieve the suffering. Union Carbide has steadfastly refused.

Settlements of less than $4,000 for families of a victim killed and as little as $1,900 for the injured have left tens of thousands of Bhopalis with no hope of a life other than utter destitution.

This stands in sharp contrast to the $4.9 million in pay and bonuses going to Union Carbide's six top executives in 1994.

But the worst news is that Bhopal may be just a warning of things to come.

If Companies Can Act Like Outlaws Overseas, Why Not In The U.S.?

Bhopal is the dark side of the "new world economy." Many U.S. companies are rushing to hungry nations powerless to enforce environmental and safety laws.

They've learned from Bhopal how much they can get away with. Cold-blooded calculation says it pays to transfer plants and jobs to countries that can't regulate. The result? Instead of poor countries upgrading their rules to match the U.S., we're being squeezed to weaken our own.

Teach These Reckless Companies A Lesson They Won't Forget.

Show the rest of the world that we won't tolerate U.S. companies abandoning us to ruin the lives of others. And show top corporate leaders that they may run, but they can't hide.

- Asia-Pacific Environment Network
- Calhoun County (Texas) Resource Watch
- Center for Community Action & Environmental Justice
- Citizens Clearinghouse for Hazardous Waste
- Communities Concerned About Corporations
- Council on International and Public Affairs
- Earth Island Institute
- Environmental Action
- Environmental Research Foundation
- Food & Water, Inc.
- Friends of the Earth U.S.
- Grassroots Policy Project
- Greenpeace USA
- Indigenous Environmental Network
- Institute for Agriculture and Trade Policy
- Kentucky Environmental Foundation
- Learning Alliance
- Military Toxics Project
- Pesticide Action Network (North American Regional Center)
- Rocky Mountain Peace Center
- Silicon Valley Toxics Coalition
- Southeast Regional Economic Justice Network
- Southerners for Economic Justice
- Southwest Network for Environmental & Economic Justice
- U.S. Urban Rural Mission (World Council of Churches)

Source: *New York Times*, 5 December 1995

The unlikely heroines (2004)

Penny Wark's article in the Times (U.K.) about Rashida Bi and Champa Devi, winners of the Goldman Environmental Prize.

The toxic legacy of the explosion of a pesticide factory in Bhopal is still felt 20 years on. Our correspondent meets two women who began a global campaign seeking justice for the victims of Union Carbide.

They are calm, these two women. They have none of the hustle of Western campaigners. Instead they tell their story with the immense dignity that comes from their culture and from the knowledge that its details are shocking enough.

Twenty years ago Rashida Bee and Champa Devi Shukla lived near the Union Carbide pesticide factory in Bhopal, central India. Champa lived 500 yards (560m) away and on the night it exploded, leaking toxic gases throughout the area, she tried to escape with her husband and their five children. Their eyes felt as if they were on fire, their throats stung. They saw bodies in the streets and piled up at the hospital. Champa still remembers the smell: of dead people, of burning flesh, of rotting animals.

Today the factory is derelict but the Bhopal tragedy continues: toxins still leak into the water supply and the soil. Babies are born deformed, children fail to grow, those who survived the night of December 2, 1984, endure chronic health problems: menstrual irregularities that lead to early menopause, fevers, tuberculosis, cancers. Champa's husband and two of her sons are dead: her eldest son killed himself in 1992 because his breathlessness meant that he couldn't work, and he became depressed. One daughter is partially paralysed; both daughters were rejected by the families into which they married because of their poor health; a granddaughter was born without an upper lip. If asked, Champa will tell you all this not because she wishes to repeat the details, but because her story is not unusual in Bhopal. She wipes her eyes frequently. She is not weeping – she is beyond tears – but the effect of the gas still makes her eyes stream.

It is for these reasons that Rashida and Champa have spent much of the past 20 years fighting for justice. They want Union Carbide, now owned by Dow Chemical Company, to be held accountable for the immediate deaths of 8,000 people, the 20,000 deaths attributed to the disaster since, and the continuing suffering of those who survive, and who have been born in the locality since. Dow should pay for the medical treatment and rehabilitation of those who have survived, and their children, and it should clean up the abandoned factory, the women insist, not unreasonably. And now their efforts have been recognized by the award of the Goldman Environmental Prize, given annually to grassroots environmental heroes. Rashida and Champa have donated their $125,000 (£70,000) prize money to a trust for Bhopali children deformed by the poisoned gas and water.

They speak of a recent decision in the Indian Supreme Court that requires the Government to supply piped water to Bhopal. "This is one of the many victories," says Rashida. It is indeed a remarkable achievement, given that before 1985 neither woman knew anything of India's political hierarchy. Both had married at 13 and, as Champa puts it, their expectations were straight forward: "Nothing other than thinking that I would have kids, and they would have kids, and I would be a grand old woman." They knew nothing of the world beyond their own communities.

Yet today they are global campaigners, visiting the US and now Britain to raise awareness of the consequences of corporate recklessness, and to ensure that a Bhopal never happens again. Champa is a tiny woman of 52, a Hindu whose metamorphosis from victim to activist began when mounting medical bills for her family made her join a training scheme for 100 gas affected women in stationery production. She had previously earned 15 rupees (20p) a day by making papads (snackbreads) but no longer had the energy for this physical work. At the stationery office she met Rashida, now 52, who had grown up in a poor, traditional Muslim family. Having lived under the rules of purdah, she had rarely stepped out of her home and felt afraid, but on the second day she stopped wearing her burka and on the third she made it inside the workplace. Her hands shook from fear and she rarely spoke. "I didn't like her because I thought she was too uptight," Champa says.

After four months' training the women were told that there was no more work; they nominated two women to represent them: Rashida because when she did speak she didn't waste words, and Champa because she was a good leader. Champa, with her limited local political knowledge, vaguely knew of the existence of a chief minister to whom they could complain. Undaunted, she and Rashida set off for his office, asking directions as they went. This was their first protest and, 19 months later, in December 1987, they set up the Bhopal Gas-Affected Women Stationery Employees Union. Demanding regular wages and better facilities, they staged a sit-in, hounded officials, and won improvements in the women's working lot.

But still they were defined as "daily wage workers", earning a sixth of the wages earned by men at other government presses. They wanted equal pay, and proper medical care and gainful employment for all Bhopal survivors. This was when they came up with their most audacious protest: putting their demands to the Prime Minister in person. They knew he was based in Delhi. "We had no money so we thought, let's walk," says Rashida. "We didn't know where it was and how far it was." Thus 80 women, 40 children and 12 men set off for Delhi in 1989. The adults took it in turn to carry the youngest children. "We had one thing in our head: to reach Delhi and talk to the Minister," says Champa. "After we had travelled 28 miles (45 km) we were asked to see the chief minister. He said: 'Why are you doing this with so many children? Why are you risking your lives?' We told him that if he would agree to our demands in writing, we would call everyone back. He refused."

This is the kind of understated and unequivocal determination that has brought the women recognition. "Sometimes when the children would not sleep because of hunger, women sold their anklets and necklaces," says Champa. "These are things you keep all your life for the well-being of your husband. But often we met generous people who would offer us food."

These were the days of rain, says Rashida. "We crossed forests with snakes and scorpions. There were ravines full of bandits; people told us that our honor would be taken from us. We didn't get hurt, and in the bandit country a local police chief accompanied us for 40 km. He cried when he left us. He said we were women but we were still brave." Rashida smiles.

It took 33 days to walk 469 miles to Delhi. There they were told that the Prime Minister was abroad and could not see them. "It taught us the nature of struggle," says Rashida. "We did not get anything there but our enthusiasm was heightened and that was a great lesson."

In 1989 Union Carbide agreed to pay $470 million in damages, but many survivors got little or nothing, and what they have received has been consumed by medical bills. Today the Sambhavna clinic, set up with donations, provides free care, which helps Rashida and Champa, who still have health problems. But such difficulties do nothing to impede their campaigning – they continue to lobby for Dow to accept liability. In 2002 the women organized a 19-day hunger strike in Delhi, demanding that Warren Anderson, the former Union Carbide chief executive, face a criminal trial in Bhopal. Last year they collected 5,000 brooms from survivors living around the factory and delivered them to Dow's headquarters in Bombay; they called it the "Beat Dow with a broom" campaign. Now they have international recognition, which they surely deserve for their dogged refusal to let poverty deny justice to the people of Bhopal.

Source: Penny Wark, *The Times*, *Times 2*, 25 May 2004

Twentieth anniversary demands of the Bhopal survivors (2004)

The following list of demands, published by the International Campaign for Justice in Bhopal (ICJB), enumerates the many problems and needs that form the basis for agitation in the coming years.

As the current owner of Union Carbide, Dow Chemical, USA must:

1) FACE TRIAL: Ensure that Union Carbide and Warren Anderson cease to abscond from the Chief Judicial Magistrate's court in Bhopal and that authorized representatives of Dow-Union Carbide face trial in the Bhopal Court.
2) PROVIDE LONG TERM HEALTH CARE: Assume responsibility for the continuing and long term health consequences among

the exposed persons and their children. This includes medical care, health monitoring and necessary research studies. The company must provide all information on the health consequences of the leaked gases.

3) CLEAN UP POISON: Clean up toxic wastes and contaminated groundwater in and around the Union Carbide factory site. Provide safe water to the community, and just compensation for those who have been injured or made ill by this contamination.
4) PROVIDE ECONOMIC AND SOCIAL SUPPORT: The Corporation must provide income opportunities to victims who cannot pursue their usual trade as a result of exposure induced illnesses, and income support to families rendered destitute due to death or incapacitation of the breadwinner of the family.

By the legal and constitutional rights of the survivors,

The Indian Government must:

1) Set up a National Commission on Bhopal with the participation of non-government doctors and scientists and representatives of survivors for long term health monitoring, research, care and rehabilitation of the survivors of the disaster and their children at least for the next thirty years.
2) Take immediate steps to send an amended request for extradition of Warren Anderson and for extradition of the authorized representative of Union Carbide Corporation. Set up a special prosecution cell in the Central Bureau of Investigation to expedite the pending criminal case against the Indian subsidiary and Indian officials of Union Carbide.
3) Ensure scientific assessment of the depth and spread of groundwater contamination and epidemiological study of exposure-induced diseases. Support the class action lawsuit filed against Union Carbide Corporation in the US Court.
4) Declare 3 December as a National Day of Mourning for the Victims of Industrial Disasters. The disaster in Bhopal must be made part of textbooks in school and university education in the country.

By the constitutional rights of the survivors,

The State Government must:

1) Supply safe drinking water through the Kolar Pipeline in communities affected by Union Carbide's contamination and contain the above and below-ground chemical wastes until their safe disposal.

2) Present a White Paper on expenditures made, programs carried out and results obtained in the last twenty years with regard to the relief and rehabilitation of the survivors.
3) Provide free treatment to residents of contamination affected communities and persons born to exposed parents and assess the health damage caused by exposure to Carbide's poisons.
4) Support survivor's attempts to build a memorial to the disaster at the site of the Union Carbide factory.

Source: www.bhopal.net

Twenty years later, a growing students' movement worldwide (2004)

Ryan Boydani, an activist who runs "students for Bhopal" summarizes student activism on Bhopal around the world on the occasion of the twentieth anniversary.

This year, students from more than 70 colleges, universities, and high schools around the world organized and participated in a wide range of protests, demonstrations, and educational events to mark the twentieth anniversary of the Bhopal disaster. These events were organized by Students for Bhopal, Association for India's Development (AID) chapters, the Campus Greens and the Environmental Justice Program of the Sierra Student Coalition (SSC), and represent the first mass student movement Dow has faced since the Vietnam War. The number and range of student events exceeded all our expectations, and represent a real and growing threat to Dow's reputation, and its ability to do business with colleges and universities across the country. At the same time, student interest in the campaign continues to grow, and students from more than a dozen additional schools have expressed interest in organizing their own Bhopal campaigns. The momentum for Bhopal among students has never been greater.

Here's a look at student events for the twentieth anniversary (a complete listing of all events can be found at http://www.studentsforbhopal.org/GDA2004.htm):

Who Participated?

- 70 schools – a huge jump from the 25 that organized events last year.
- 22 AID chapters
- 28 US schools held their first Bhopal event
- 8 international schools, plus several more from the Philippines, held their first Bhopal event.

Where Events Took Place?

- 57 schools held events in the US (in 23 states)
- 14 schools held events internationally

What Events Were Organized?

- 42 documentary screenings
- 14 vigils
- 13 information tables/flyerings
- 11 petition signings
- 9 speaking engagements or panel discussions
- 8 Bhopal Express screenings
- 8 school newspapers/magazines printed articles written by students in the campaign (others were also printed, but are not counted here)
- 7 protests against Dow Chemical
- 6 photo exhibits (not counting those in the Philippines)
- 5 universities face major efforts by students to end university associations with Dow.
- 5 classroom presentations, 4 posterings
- 3 hunger strikes
- 3 essay or drawing competitions
- 3 performances/plays
- 3 slide shows
- 2 public debates
- 2 letter-writing efforts to Congress
- 2 fundraisers
- 2 banners/petitions
- 1 fact finding report
- 1 student-produced film release
- 1 conference
- 1 human chain

Highlights

Chicago, where 24 activists from the South Asian Progressive Action Collective and allied organizations including AID Milwaukee delivered *jhadoos* (brooms) to Dow Board member, James Ringler. Ringler answered the door, ordered the activists off his property, and hastily closed it, refusing to accept the brooms given along with a letter and the recently published book, "Trespass Against Us":

"We started by reading a survivor testimonial aloud and then read aloud our open letter to James Ringler. Two folks went up to the door to present Ringler with a *jhadoo* and letter which he refused and told us to get off his property. We left the *jhadoo,* the book "Trespass Against Us" and the letter on his doorstep and then went back out to meet our group. We stayed on the street in front of his house and shouted chants and marched so that he would know for sure that this issue was not going away. We were honored to have a visiting delegate from the New Trade Union Initiative in India with us, who is here as part of a collaborative exchange/ dialogue with Jobs with Justice so that Indian workers and American workers can start building relationships and organizing to prevent any such disaster from ever occurring again. The cops came but did not chase us off and actually took our fact sheets and just waited while we did our thing. All in all a very positive night!"

Washington, where supporters of ICJB paid a visit to Dow board member Barbara Hackman Franklin's office to deliver a letter from the group, "Trespass Against Us: Dow Chemical and the Toxic Century" and a vial of contaminated Bhopal well water.

Delhi University in India, where the members of the student group "We for Bhopal" released the report of their October 2004 Fact Finding Mission to Bhopal, for which students met with survivors, toured the factory grounds, and interviewed the Chief Minister of the state government and other officials. The students intend to deliver the report in person to the President and Prime Minister of India, following up on their meeting with the President in March. In addition, "We for Bhopal" organized a massive candlelight vigil to mark the anniversary that drew a great deal of press attention, and judged the results of its college essay competition.

University of Texas, Austin, where the members of AID-Austin organized a three-day-long series of events. These included a day-long protest against University involvement with Dow, a hunger strike and candlelight vigil, and a film screening and open discussion about the disaster.

St. Benedict's Preparatory High School in Newark, New Jersey, where the members of the *SBP Environmental Club* showed the video "Twenty Years Without Justice" and a slide show to the 650-strong student body. A copy of the film and the book "Trespass Against Us" were donated to their high school library. On Friday 3 December, a group of students went to NJIT and Rutgers and handed out information slips. "In total we got 650 students and 100 teachers at St. Benedict's and at NJIT and Rutgers we got to speak with about 150 students and about 50 teachers."

Source: www.studentsforbhopal.org.

11

Evasion

Getting away with murder

ACROSS TWENTY SPIN-FILLED years, Union Carbide's treatment of Bhopal has been a stellar example of evasion through creative downsizing and ingenious public relations. Rather than responding proactively and openly to the unprecedented chemical accident, Union Carbide has chosen to conduct business as usual: unremitting denial, obfuscation and manipulation. The stated position of the corporation on the Bhopal tragedy is that it was an act of employee sabotage that caused the accident. Detailed investigations into the precise causes of the incident, including theirs, are reprinted in this text (see the chapter, "Explanations"). However, beyond the implausibility of the sabotage theory (Carbide's own lawyers abandon it when convenient, as seen in Claude Alvares' report below), the fact remains that employee sabotage is a foreseeable danger and is neither a moral nor a legal defense in the Bhopal case.

This chapter traces the corporate version of the Bhopal story – from Warren Anderson's launching of the "sabotage theory" at a press conference in March 1985, through Dow Chemical's negotiation of the issue today. The statements of generations of Carbide and Dow CEOs and spokesmen are represented here, but as the article "Face to Face Encounters with Carbide Officials" demonstrates, the lawyer whispering in the spokesperson's ear is really the only voice there is. It's not personal; it's public relations. The corporate entity's unflappable devotion to the bottom line, combined with the particularly inflexible, reactionary, and perhaps discriminatory culture within Union Carbide specifically (a description of which is provided below by Wil Lepkowski, a chemical industry based journalist), have meant that there has been, in more ways than one, no "human" reaction to the disaster.

Compounding the continual denial of responsibility has been a consistently petty and miserly attitude towards the issue and the survivors – a policy of taking credit for as much as possible, while giving as little as possible. While they depend on listing the different amounts of money that they have given or "attempted" to give – as though they get extra credit – the numbers, which aren't much to begin with, add up to little more than creative accounting (see Sprick to Hanna, 2004, 2005 below). For example, when the Indian Red Cross initially accepted $5 million from UCC, they specifically stipulated in court that this money should not be offset against the final settlement. However, when Union Carbide did settle with the Indian Government, they refused to pay the full amount of $470 million dollars, claiming that they now owed only $465 million. But since the $5 million could not be offset, they did the unthinkable; the court ordered that the Red Cross *would return the money to the settlement fund set up by UCC.* The Red Cross hospitals were closed (these hospitals were discussed by Verweij in the chapter, "Healing"). In any case, Sprick maintains that "UC does not know how the monies were used, who benefited and how it may have ultimately been distributed." In the end, of the many beneficent gestures UCC proudly credits to itself, the settlement itself is basically the only one that actually made its way to the survivors (such as it did – see chapter, Just Compensation? for more about that money). But that's not Union Carbide's problem – what's important is the paper numbers for the people who read about Bhopal on paper and on-line, not the real, messy business of "who benefited" from these abortive projects.

In 2001 Dow Chemical assured its shareholders that Carbide was a safe and lucrative acquisition, but Carbide has turned out to be much more of a liability than anticipated. Although the disappearance of Carbide into Dow was the dénouement of a long-term policy that has made the company invulnerable to prosecution through divestment and name changes (see for example, the layers of corporate renaming that, Dow India argues, keeps them from presenting Union Carbide, at the end of the chapter, "Criminals – and Corporations"), it has also provided activists with a much larger and higher profile target. Bhopal is now Dow's burden – and if their recent retention of the notorious PR firm Brown, Nelson & Associates (see excerpt from their promotional literature below) is any indication, it's one they take quite seriously.

In fact, from international protests to shareholders' resolutions, Bhopal has proved to be an onerous enough liability that Dow has begun creating artificial distinctions to dissociate itself internally from its fully owned subsidiary. For example, while Dow monitors listservs and websites about Bhopal (as shown in the last excerpt in this chapter, from spokesmen John Musser and Tomm Sprick), when making any written statements Dow does so under the moniker of the Union Carbide Corporation "since Dow has no responsibility for the Bhopal disaster" (Sprick to Hanna, 2004). In the fall of 2004, several years after they acquired Carbide, www.unioncarbide.com suddenly appeared. This site stages Carbide's "independent" version of the Bhopal disaster, thinly veiled as a functioning corporate website (for more information on how the Bhopal story plays out on-line, see the following chapter – "Responsibility").

This arbitrary distinction is quintessential to Dow's newest strategy on Bhopal. Dow makes a symbolic distinction between what has been done and known by Union Carbide, and what is done or known by Dow. Pressed for information, on, for example, the sabotage theory, Dow's response is that they don't know what happened, but Union Carbide says that it was a saboteur, and Dow believes them – expressly without having done any research of their own. This chapter presents these two corporate positions on Bhopal in their own very expensive words.

Remarks by Warren Anderson, March 20 (1985)

Carbide's CEO at the time, Warren Anderson, at a press conference in March 1985, began trying to shift the blame to "operating plant personnel" – i.e., workers employed by Carbide's Indian subsidiary and to hint at the possibility of worker sabotage.

We believe the lessons to be learned from a tragedy like Bhopal are too important to be based on speculation and conjecture. We believe they deserve the careful study and analysis provided by our investigative team and all the scientific resources available to it.

We can say with a great deal of confidence what went wrong technically at Bhopal. In addition to the deliberate or inadvertent introduction of a large amount of water into tank 610 as the report describes, the investigation team has identified a number of independent operating events and circumstances which taken together all contributed to the incident in Bhopal. Those events and circumstances, which Ron Van Mynen [Corporate Director of Health and Safety] will cover in his report, were not in compliance with standard operating procedures. Compliance with these procedures is the responsibility of operating plant personnel.

The culture and mind of Union Carbide (1994)

Chemical industry based journalist, Wil Lepkowski, explores the internal culture of the Carbide Corporation.

The culture of Union Carbide

Assembling a fully descriptive portrait of Union Carbide through its transformation since 1984 is an elusive undertaking (see my postscript). That is because, as with all corporations, at least four worlds intersect within its closed institutional system, each painting its own canvas. First is the financial realm, demarcated by the company's performance on Wall Street; second, the complex managerial world of covert strategies, veiled motives, and risky decisions; third, the human world of the Carbide work force; and, fourth, the symbolic world – the sturdy and reliable image that the corporation seeks to present to its customers, its shareholders, and the public.

Sharply different, too, are the confident, masculine, all-is-well, all-will-be-well mentalities at the headquarters level in Danbury, Connecticut, and truer-to-life existence of smells, and Louisiana, of Carbide the grittier, sounds, laughter, emergencies at Carbide's operating plants in West Virginia, Texas, and elsewhere. Carbide's saga is replete with drama and much known, a great deal hidden – because of the legal, financial and media pressures that bore down on it from late 1984 through 1989, when the case reached the

Indian Supreme Court. It survived, and bears its combat medals proudly – though as regards Bhopal rather stiffly, often evasively.

So the post-Bhopal Carbide brought into rough focus in this chapter is no anatomically correct rendering of a complex organization but instead a shifting cubist melange. The parts are recognizable but they frequently seem oddly disconnected. The legal Carbide hardly fits an outside observer's expectation of the ethical Carbide, though Carbide officials would clearly dispute that assessment. The scientific Carbide – especially as regards toxicological research on methyl isocyanate – would hardly live up to science's exacting demand for all the toxicological facts about a controversial molecule. And the journalist's ideal of a Carbide candid about itself, willing to take a reporter aside and describe the agonies of decision, conflicts rather predictably with the reality of a corporation bent on preserving shareholder value by protecting its image.

Historically, Union Carbide's corporate structure likewise displays a cubism of its own behind the company's hexagonal logo. Separate, unrelated fiefdoms of industrial gases, metals and alloys, carbon electrodes, agricultural products, chemicals and plastics, and specialty products each jealously adhered to their own corporate personas, presenting an image of a company perpetually searching for focus. New ventures would emerge according to fashions of the managerial moment, only to flounder in indecision and isolation, and ultimately disappear. To Wall Street, the pre-Bhopal Carbide seemed a benign, inbred, poorly managed, under-performing, company that never could quite "get its act together," an image that dogged the company from the mid-1960s until its systematic dismemberment into the monolithic ethylene-based petrochemical company it returned to by 1993.(1)

What all Carbiders have always held in common, though, was pride in technology and the feeling of being treated fairly and humanely by the company. Carbide's top managers, in my judgment, displayed the once stereotypical mentalities of the engineer: regimented, narrowly focused, business-first, hostile to the social and political values of the environmental community, but paragons of dedicated citizenship.(2)

Until the late 1980s, when continuous restructuring and layoffs led to serious declines in the morale of its workers and depressed confidence in management, (3) Union Carbide commanded extreme employee loyalty and security. Though less than a sixth the size it was before Bhopal, Carbide still ranks as a large chemical company, employing around 14,000 people in plants along the Houston Ship Channel, the petrochemical alley running down from Baton Rouge to New Orleans in Louisiana, the Kanawha Valley of West Virginia, and dozens of other places.

In the Kanawha Valley, many Carbide families reach back to the company's beginnings in 1917. Bonds are powerful among Carbiders, as well as between Carbiders and the various social and commercial institutions in the Valley. To be a Carbider – at least to be one still remaining with the company – is to wear a badge of reliability, responsibility, and stability. When reminded of the Bhopal disaster, a Carbider always replies that the cause was sabotage at a plant operated by Third World foreigners. The American company could not possibly have been responsible. (4)

Before Bhopal, "the whole chemical industry operated on the basic assumption that what we did within our fences was none of anyone's business," commented one Carbide community relations official in Charleston interviewed in 1991. "And the people outside the fences didn't think it was any business of theirs, either. Bhopal changed all that, and for the better." Yet, ironically, what really brought new consciousness to Carbide, he added, was not so much Bhopal as the serious leak of a mixture of aldicarb oxime that took place nine months later at its plant in Institute, West Virginia. "That one really shook us," he said. "We said it couldn't happen to us here and it did. That's what really kicked in these new attitudes." (5)

But Bhopal remains central to Carbide's damaged identity. The noble mythology of the chemical calling (high service to society) made the enormity of the disaster especially jarring to Carbiders. It was their children who had to face classmates in the disaster's aftermath. Carbiders felt not only the shame of identification with the affected company but also the insecurity of wondering about the survivability of their jobs. And those fears were well placed. As a result of the corporate dismemberment during the 1980s and early 1990s, tens of thousands of former Carbiders were either let go or with some degree of relief became part of such companies as Ralston Purina, Rhone-Poulenc, First Brands, UCAR Carbon (the name of a new company formed by the sale of half of Carbide's carbon products division to Mitsubishi), and Praxair. This dissection almost certainly could have been avoided had Carbide made an early $600 million settlement on damages with the Indian government.

Up to the editing of this chapter in mid-1993, Carbide was well embarked on a campaign to demonstrate to all its publics its determination to be "second to none" in environmental and safety matters. But the effort of winning over the public has been a difficult one for Carbide, owing at least partially to its past. Of no help was the 1985 aldicarb oxime leak, or the March 1991 explosion of an ethylene oxide unit at its petrochemical plant in Seadrift, Texas, which killed one person and injured thirty-two. For that accident, the Occupation Safety and Health Administration fined Carbide $2.8 million, reprimanded Carbide for ignoring several internal safety audits that urged preventive measures in the explosion area, and withdrew a safety award the agency had given Carbide months before. The report conjectured that had the accident occurred during the workday, rather than at 1:00 a.m., more than four hundred might have been killed. (6) As a result of the Seadrift explosion, a group of community activists formed an organization they called "Citizens Concerned About Carbide" in response to what they felt was Carbide's less-than-forthcoming behavior within the surrounding community following the Seadrift explosion. By the end of the year, they were petitioning the Chemical Manufacturers Association (CMA) for violating the principles of its Responsible Care accident prevention program. The effort failed, but their actions resulted in the launching of a continuing dialogue between the two sides over the sharing of relevant safety information between the plant and the community.

I have dwelt at length on this background in order to say simply that attitudes at Carbide as a company are representative, even stereotypical, of traditional chemical industry attitudes in America before and since Bhopal. Before Bhopal, it distrusted environmentalists with lightly veiled contempt. Now, though inbred defensive attitudes persist, it recognizes the tides and even characterizes itself as a "green" company. (7) Carbide's attempt to locate two small chemical plants in Kingman, Arizona against strong public opposition offers one illustration of its new attitudes. (8) Because two chemically unrelated plants were involved, entailing different aspects of Arizona's environmental legislation, the action aroused a swirl of corporate and citizen misperceptions. But in the end Carbide learned important lessons of community relations that were not lost on corporate headquarters. (9) Carbide succeeded in building the plants through considerable compromises with the community, through almost unprecedented negotiation between plant managers and citizen groups, and at a price that barred any further chemical development in Kingman. At Seadrift it was learning, with much more difficulty, to deepen its exposure to the local public. For what community activists were pressing for was in essence internal co-management of the plant's accident-prevention system by establishing an inspection corps made up of outside experts.

As Carbide's then-president, William Lichtenberger, told shareholders during the company's annual meeting in April 1991, the high costs of environmental control and safety engineering are "absolutely necessary to meet our responsibilities to our people, our neighbors, and to the planet." It was only when the 1990s began, six years after Bhopal, that high-level Carbiders began so explicitly to express the company's responsibility to the earth's ecology and to employ such verbal icons as "planet" in their rhetoric.

The Mind of Union Carbide

Is there any final summation, any accurate generalization, of the "mind" of Union Carbide as reflected in its behavior during the eight years since Bhopal? Answers must be cautious because they are personal and are mostly derived from inference.

Carbide can justly claim to have been misrepresented and unfairly abused in its attempts to bring assistance to the Bhopal victims. Self-righteousness certainly pervaded all sides – Carbide, the Indian government, and activist groups – during the entire drama. Industrialists may have a heart as individuals, but their professional devotion and sense of survival cling tenaciously to the financial bottom line. To have admitted any mistakes in managing the relationship with its Indian affiliate would, to Carbide officials, have spelled destruction to their company. Yet corporate officials do agonize, do stumble, and do get muddled in the process of establishing policy. When in 1984 I asked one Carbide officer how he would have managed UCIL, he answered simply and probably accurately: "I would have kicked ass."

Carbide worked hard to portray itself as a passive victim of foreign-investment laws and rules in India. Other companies ridiculed this stance, but, then, none of them suffered a Bhopal. According to its own account, Carbide also became a victim of India's radical politics and technological backwardness and finally of an alleged saboteur. Beginning as the party "morally responsible" for the disaster, Carbide painted itself finally with continuous brush strokes as the moral victor. And Carbide did not fail to receive its share of support from the financial and popular press. (10) Especially supportive was a 1988 broadcast of the CBS program "60 Minutes" that sympathetically described Carbide's frequently thwarted attempts to bring humanitarian aid to the victims.

Still Carbide's image during the Bhopal episode was that of a company that had decided to play legal hardball with the disaster. While it claimed moral responsibility, it seemed to shun any deeper spirit of reparation or of understanding Indian culture. How can one explain that? There is a form of dishonesty, or perhaps more properly structural self-deception, built into the process of corporate reparation in an industrial disaster. Such a posture (based on the implicit proposition that "we do everything well, carefully, caringly, and safely") may be unavoidable, because liability is always just around the corner in any chemical operation. But it exists nonetheless, supported by two kinds of institutional pressure. The first is the need to put only the best face forward to shareholders, present and future. The second is the unavoidably litigation-resisting character of the modern U.S.

corporation, which translates into the position: "We can make no mistakes that can be admitted to."

In the Bhopal case, the first imperative led to Carbide's attitudinal "atonement" through its establishment of new safety and environmental practices within the company. Bhopal drove Carbide to do good for its own workers, as it also led to the still deficient community right-to-know law in whose promotion Carbide played a leading role. At the same time, the second imperative pushed Carbide to distance itself from the disaster victims in Bhopal. While it "atoned," it also detached.

Union Carbide, in my view, had every reason to reach out for public understanding – to more fully atone, as it were, for an accident that occurred under its blue and white logo in a distant part of the world. Warren Anderson spoke of Carbide's "moral responsibility," and it seemed just the right tone to take in those early days. But it raised false expectations that Carbide would seek creative solutions as the legal picture became more and more tangled and contentious. Later, Carbide transmuted "moral responsibility" into self-righteousness, becoming moralistic, even close-minded, about Bhopal. To Carbide, the settlement was a closure that allowed it to walk away from India, to evade the fuller atonement that moral responsibility implies. Bhopal could have been the opportunity for Union Carbide to display legal and moral innovation: a disaster one company decided not to turn its back on. Instead, it negotiated not a commitment to continuing stewardship at Bhopal, but an uncreative, even antiquarian, way of notarizing its moral responsibility for what was (and is) a unique, ongoing tragedy.

In the summer of 1989, during a long interview at Washington's Metropolitan Club and later in his office, I asked Ronald Wishart whether Carbide bore any long-term responsibility to atone for Bhopal. The answer given by Wishart was that atonement implied guilt, which Carbide never intended to admit. But Wishart misread his Christian theology. The idea of atonement in the Christian tradition is not primarily about the atoner's guilt; it is about sacrifice as an act of reparation for the sins of others. By its silence, its self-righteousness, its lack of scientific integrity in failing to pursue the toxicology of methyl isocyanate, and its refusal to bring to justice any Bhopal plant "saboteur," Carbide shunned the ideal of corporate moral responsibility, while publicly embracing the concept. (11)

In 1989 Carbide's chairman, Robert Kennedy, was elected chairman of the Chemical Manufacturers Association (CMA). The theme of his speech before CMA's 1989 annual meeting was the chemical industry challenge of regaining the "trust and confidence of the public." He said: "We can sit back and wait for the public to come to its senses and discover us doing good, or we can stand up strong and tall and say what we're doing, what we've done already and what we're going to do. There is a growing need for predictable, consistent environmental stewardship from nation to nation and region to region around the world. The chemical industry has a great opportunity in the years ahead to make new history."

Eight days after Kennedy's speech, an internal memo was distributed among Carbide's public affairs strategy group. It concerned an environmental activist organization known as the Citizens' Clearing House for Hazardous Waste (CCHHW) formed by former Love Canal activist Lois Gibbs. The memo characterized CCHHW as "one of the most radical coalitions operating under the environmentalist banner," with "ties into labor, the communist party, and all manner of folk with private/single agenda." CCHHW's agenda, the memo said, was to "restructure U.S. society into something unrecognizable and probably unworkable. It's a tour de force of public issues to be affecting business in years to come."

Carbide officials discount the memo's significance. What it did do was expose a lingering mentality the company finds difficult to shake, even after Bhopal – that "radical" elements of society somehow conspire against the virtues embodied by American industry and, perforce, against civilization itself. But by Earth Day 1990, just as the *Washington Post* was about to publish the text of the memo, its author sent an apology to Gibbs. That gesture could also be construed as an offer of reconciliation, another step in Carbide's long voyage toward renewed respectability.

Union Carbide after Bhopal is a permanently changed company. Its officers may personally have profited by the breakup of the company after the GAF takeover attempt, but it is deeply scarred and permanently pained by the experience. Kennedy won for himself and Carbide their self described "fistful" of awards from the chemical industry. But in the post-Bhopal era, this whole industry is in the process of transformation toward sensitivity and gentility in planetary stewardship. Belatedly, Carbide, too, seems to have learned the right vocabulary and has, by all indications, finally "got it."

At the 1991 annual meeting, Carbide's legal counsel and director, Joseph Geoghan, would tell me: "Bhopal is going to be the kind of overhang that will always drive Union Carbide and the industry toward improved results. Everything we do will be put in the context of Bhopal." Yet by 1992 Carbide seemed to be refocusing itself into oblivion. Entities bought only a few months before were being offered for sale. All that was left of the Union Carbide that before Bhopal had employed almost 100,000 people was its Chemicals and Plastics core, with a smattering of unwanted specialty operations and a work force only 14,000. Kennedy was described as revivified at the prospect of running a company now so tightly focused. Yet in 1993 UCC continued to struggle for profitability. The world market in ethylene-based chemical commodities remained in a deep recession. Added foreign capacity intensified the competitive pressures on Carbide. Prospects were grim.

As a boy during the 1940s, I remember attending frequent movie matinees that would often run cartoons featuring the characters Bugs Bunny and Porky Pig. Each feature would close with Porky focused at the center of a large circle. Before "The End" appeared, and while the band played its merry tune, the circle would diminish in size to a narrower and narrower focus before disappearing entirely, taking Porky with it. As the circle narrowed, Porky would intone his good-bye to the audience with his cheery, "That's All, Folks." And then, Porky gone, the theater screen would darken to silence. So it may be with Carbide, except that the story is hardly a funny cartoon, and there is no applause.

Footnotes:

1. This judgment is based on informal conversations with former Carbiders and others familiar with the chemical industry. Carbide, however, has been considerably written about in the financial press. See, for example, *Forbes* (December 10, 1990): 106; *Business Week* (October 29, 1990): 70; and *Wall Street Journal*, January 23, 1992, p. A1, for critical accounts of Carbide's corporate prospects.
2. Corporations are careful to remain "off the record" in their honest attitudes toward the environmental community. This judgment, however, is based on my own years of reporting on the environmental movement, including involvement in several off-the-cuff conversations with industrialists over the years.
3. An internal survey of Carbide employees, conducted late in 1991 and published in its company magazine, reported a widespread decline in confidence toward top management. See Union Carbide World (December 1991).
4. These assessments are self-evident to anyone who has spent any time living in or visiting a chemical plant community. The chemical industry culture has indeed not been studied to any extent. But in my own visits to the Kanawha Valley and through conversations there, I repeatedly was struck by the strong sense of honesty and integrity of Carbide employees, along with a thin-skinned readiness to defend the vulnerable record of the company. Carbide, after all, was also responsible for the worst industrial disaster in American history – the infamous mass silicosis tragedy during the construction of the Gawley tunnel during the 1930s by Carbide's metals division. See Martin Cherniack, *The Hawk's Nest Incident* (New Haven, CT: Yale University Press, 1986). Carbide omits that episode from its histories.
5. Interview with Thad Epps, director of community affairs, UCC, South Charleston, WV, July 1991.

6. Memorandum dated June 28, 1991, "Request for consideration of approval of an egregious case," p. 4. Confidential internal report by the Occupational Safety and Health Administration, 1992.
7. Cornelius C. Smith (UCC vice-president, community and employee health, safety, and environmental protection), "Bhopal Aftermath: Union Carbide Rethinks Safety," *Business and Society Review* 75 (Fall 1990): 50.
8. *Chemical & Engineering News* (January 1, 1991): 15.
9. *Chemical & Engineering News* (May 13, 1991): 14
10. See, for example, *Barrons,* February 1988, lead editorial.
11. According to much Western Christian tradition (using biblical sources such as Isaiah 53: 4 – 5 and Mark 10: 45), the act of atonement in the sacrifice of Christ delivers all persons out of bondage to their sins. A self-emptying, reconciling act, the atonement does not simply cancel debts up to the present, but also covers all future debts. Belief in this theory is in turn inherent in the ethical value system in the West. Union Carbide assumed the mantle of "moral responsibility" for the Bhopal disaster, thus inviting a deeper examination of what such a claim does indeed represent for corporate human beings in a disaster on the scale of Bhopal.

Source: Will Lepkowski, excerpted from Sheila Jasanoff, ed., *Learning from Disaster: Risk Management after Bhopal,* Philadelphia: University of Pennsylvania Press, 1994.

Public relations and sabotage (1994)

In response to Ward Morehouse's review of Jamie Cassels' book The Uncertain Promise of Law: Lessons from Bhopal *(July/August 1994),* The Ecologist *received the following letter from Bud Geo Holman of the law firm Kelley Drye & Warren, and Morehouse's response:*

Bud Geo Holman writes:

Let me set the record straight that Union Carbide, its lawyers and lawyers from the New York-based firm that assists Carbide on Bhopal litigation (Kelley Drye& Warren) continue to be absolutely convinced that the terrible tragedy resulted from the deliberate actions of a disgruntled Indian employee at the plant. No Carbide or Kelley Drye & Warren lawyer ever told a Yale Law School audience otherwise. I gave a talk at Yale and made it clear that employee sabotage was, in fact, factually and legally well based. For Mr. Morehouse to state the opposite is outrageous and false. The only public relations ploys that deserve to be questioned are those that he and the so-called International Coalition for Justice in Bhopal have perpetuated for nearly 10 years.

Ward Morehouse writes in reply:

At a session of the Yale Forum on International Law on 14 December 1989, I heard Paul Doyle, Mr. Holman's colleague at Kelley Drye & Warren, concede under sharp questioning (not by myself) that what Mr. Holman calls "the deliberate actions of a disgruntled Indian employee" would not have any significant bearing on the legal determination of liability. If the sabotage theory had no legal merit, what was its purpose other than as a public relations ploy to make Carbide appear the victim, not the victimizer, and to divert attention from the real causes of the disaster in the careless design and operation of an extremely hazardous chemical facility?

Mr. Holman claims the sabotage theory is "factually and legally well-based." But according to Carbide's own safety manual, it would have taken 23 hours for water deliberately introduced in the storage tank containing methyl isocyanate to produce a violent, runaway reaction. Yet this reaction had already started at the time Carbide says the water was added and was spewing lethal gases into the atmosphere and over the sleeping city of Bhopal an hour and a half later.

This and other evidence which undermines the factual basis of the sabotage theory, including a denial by the alleged saboteur who says Carbide framed him is published in *Bhopal: The Inside Story* by T. R. Chouhan and other former Carbide workers. In the book the workers, the only direct eyewitnesses to the actual event, tell their side of the story.

Holman knows full well that there are only two defenses against strict liability: "act of god" and "act of stranger." Neither applies to the Bhopal disaster. It is well established under case law that the act of an employee cannot be construed as an act of a stranger. Furthermore, if Carbide's sabotage theory were to be taken seriously as the real explanation of what actually happened, Carbide would become an accomplice after the fact in helping to conceal a felon by not revealing the name of the "disgruntled employee" to the relevant authorities.

T. R. Chouhan writes: "The most compelling evidence of the false claims of the sabotage theory is its abandonment by the officials of Carbide's Indian subsidiary in the Bhopal Court which is hearing charges of culpable homicide against them for their role in the Bhopal disaster. Rajendra Singh their chief lawyer informed the court that the water had indeed entered tank E610 through the Relief Valve Vent Header (RVVH) as a consequence of the water washing being undertaken by plant personnel instructed to do so by their superiors. This explanation means that the water was not inducted in the tank by a water hose connected to the tank by a single disgruntled worker as the sabotage theory alleges."

Source: *The Ecologist,* September/October 1994 (Vol. 24, No. 5).

Letters to the editor: Black and white views of the Bhopal disaster (1994)

Letters published in The Nation *between Carbide public relations official Bob Berzok and Joshua Karliner, the author of an article, on the tenth anniversary of the disaster.*

Bhopal: 'Just and Reasonable'

Danbury, Conn

Nothing is further from the truth than Joshua Karliner's headline, "Bhopal – Ten Years Later. To Union Carbide, Life is Cheap" [Dec. 12]. The $470 million settlement, which Karliner says is "paltry", was settled at the direction of the Supreme Court of India in 1989. During the next few years, after the court reviewed all of the U.S. and Indian court filings, applicable law and the needs of the victims, it concluded that the settlement was "just, equitable and reasonable."

Sadly, Karliner's article also misrepresented progress made by Union Carbide and the chemical industry in their environmental endeavors in the ten years following the tragedy. Had the author taken the time to solicit reputable, independent environmental policy analysts instead of Ward Morehouse, who represents the extreme edge of the activist spectrum, he would have found that generally there is widespread support for the efforts taken to improve the safety of operation and further the dialogues with people, communities and legislators in the interests of environmental responsibility.

Union Carbide and the chemical industry led an initiative called Responsible Care, to produce distribute and dispose of chemical products safely and to work with people in communities, governments, public interest groups and activists. A lot has been done in the past decade and, certainly, there's more to be done now and in the years ahead. Saying that to Union Carbide, life is cheap, is absolutely fake, a sham and a cheap shot at the many dedicated people in Union Carbide and other companies who are working so hard to achieve industrial progress and environmental safety.

Robert M. Berzok
Union Carbide Corporation

Karliner replies

San Francisco

The "Life is Cheap" headline certainly summarizes the views of many of the thousands of victims of the Bhopal tragedy who were chagrined by

Carbide's miserly Bhopal settlement. Moreover, the Supreme Court decision – widely criticized as selling out the victims for a pittance – was in both the Indian government's and Union Carbide's interest: while Carbide's stock jumped $2 a share on the New York Stock Exchange the day of the settlement.

With regard to safety issues: As my article states, Carbide and corporate America have made some positive strides toward safer production. Yet we cannot rely on the self-regulating approach of Responsible Care to usher in a new era of ecological sustainability in the chemical industry – Carbide's ongoing problem and the industry's resistance to fundamental change make this clear. Furthermore, in this age of globalization, Responsible Care doesn't even apply to foreign subsidiaries of member companies, a glaring omission considering the Bhopal debacle.

Finally, Morehouse, a venerable and dedicated activist who for ten years has doggedly pursued Union Carbide seeking Justice for the Bhopal victims, may seem extreme to Berzok, whose unenviable job is to make his corporation look good, but to many in India and the United States, Morehouse and people like him are American heroes.

Joshua Karliner

Source: Joshua Karliner and Bob Berzok Letters to *The Nation,* January 23, 1995, p. 78,

UCIL officials abandon "sabotage theory" (1994)

Claude Alvares, an Indian journalist, tells the story of how a lawyer for the officials of Carbide's Indian subsidiary abandoned the "sabotage theory" in the Magistrate's Court in Bhopal.

The UCIL counsel, Rajendra Singh, concentrated on proving to the court that none of the nine accused led by Keshav Mahindra (Chairman, UCIL) had, with any knowledge or intention, caused the flow of water into the MIC tank. Neither, he told the court, had any of the accused anything to do with the erection of the plant – it was constructed under the supervision of experts for Union Carbide, USA.

But the most significant part of his argument, however, was his claim that the water entering tank No. 610 (from which the gas leaked) could have been due to "a rash act of negligence" by operators Rahman Khan and Salim who were both working in the plant on the night of the disaster.

Whatever Union Carbide may have wanted to hold in the case, UCIL at least was no longer relying on the claim that the disaster was due to an "act of sabotage"! According to a report of the trial in *The Pioneer* (27 Feb.), defence counsel informed the court "that the water did indeed enter tank 610 through the Relief Valve Vent Header (RVVH) as a consequence of the water washing being undertaken by plant personnel instructed to do so by their superiors."

Source: From T. R. Chouhan, *et al.*, *Bhopal: The Inside Story*, revised edition, New York: The Apex Press and Mapusa, Goa: Other India Press, 1994, (republished 2004).

Face-to-face encounters with Carbide officials (2001)

Activist Satinath Sarangi tells stories of discussing the issue of Bhopal in person – from jail cells and in conference rooms – with the individuals whose job it was to communicate Union Carbide's positions to the public.

Bob Berzok is your regular corporate executive. He wears a three piece suit and carries lines on his face that appear to come from libido worries. The first and only time I saw him – the Director of Public Relations for Union Carbide Corporation (UCC) – was through the plexi-glass screen at the Texas state prison in Houston. It was the middle of the night and from my freezing cell, I was brought into the visiting room. His company had probably realized that keeping us (two gas victims and myself) in jail on charges of criminal trespass would make bad press and sully its public image further. Bob was scared of bad PR. It wasn't even clear that we had been properly arrested; on the previous morning, officers at the annual shareholders' meeting arrested us for the simple act of distributing a fact sheet on Bhopal.

"We are concerned about your suffering," said Bob, "I have the bail money with me and a limousine waiting for you outside." "If your company is really concerned about human suffering," I said into the phone that connected us through the plexi-glass screen, looking into his eyes, "you would release all the medical information you have on the gases that leaked and are killing people in Bhopal to this day." I was watching for a change in his expression as he listened to my alien accent. Nothing happened. He repeated in words and tone exactly what he had just said. Where you and I have eyes, he had frozen cubes. He politely wished me a good night, and left.

Two years later in 1991, with my friend T. R. Chouhan – a former plant operator in Bhopal Union Carbide – I met with Joseph Geoghen in New York city. Again a regular senior executive, and a Vice-President at Union Carbide, USA. He had a lawyer sitting on either side of him and a secretary taking notes. To Joe I repeated the same request I had made to Bob.

"Your company is the inventor of industrial production of methyl isocyanate (MIC), one of the gases that leaked in Bhopal. You have been doing research on MIC and other chemicals and their effect on life systems for at least 30 years at your labs in Research Triangle Park, Raleigh. There is mention of at least 16 research studies that you have chosen not to publish, at least one of which is on 'human volunteers'. It does not cost you to give us the information you have generated over the years and for all one knows, this information may be vital in finding the painfully elusive 'proper line of treatment' for those exposed."

While the lawyers whispered behind Joe, and he waited for their conclusion, Chouhan described for their benefit, the regularity with which samples of workers' blood, urine and other substances had been taken at the factory. These reports were never released. The lawyers had finished their discussion by then and one of them, the Indian guy, whispered into Joe's left ear. We got our answer: he advised us to contact the Environmental Protection Agency (EPA) for MSDS data sheets and to take our grievances to the Indian Government.

In as few words as possible, (because I could see he was getting impatient) I described the pain of a family from my neighborhood in Bhopal, who had not known one day's respite from exposure-related illnesses and despite their extreme poverty had left no doctor or hospital unvisited. One of the lawyers indicated he had to catch a flight. I looked at Joe, Joe was looking at his watch. I knew I had little time. I appealed to him not to invoke the Trade Secrets Act and attempt to justify their continued and deliberate withholding of medical information. As politely as I could I reminded him that their denial of information was compounding the injuries they had caused – not only were they impeding development of specific therapies but they were also the direct cause of doctors prescribing drugs that were doing serious damage to peoples' bodies.

When I think of the disaster and try to fathom the minds that decided that it was right and proper to produce one of the most toxic chemicals in the midst of populated settlements; to under-design the factory that would produce that chemical, and then, to direct a global "economy drive" that, among other things, resulted in the shutting down of the refrigeration plant (to save Rs.700 per day) I draw a blank. In my generous moments I can see them just doing a job to send their children to the right school, have their wives look good at parties and keep up on the golf course. They didn't really know that a mega disaster would result from routine decisions taken as part of normal corporate practice, that of making a bigger profit than last year.

But when I think of the medical disaster that followed and is likely to continue for as long as you and I are alive, I have no generous way of thinking of the regular guys who are the principal authors of this tragedy. From the little that I know there was only one published paper on the health effects of MIC before the disaster. The only available information was held by Union Carbide. They knew – and possibly know more now – about what MIC does in the acute and the chronic phase. They know what it does to the lungs, to the eyes, to the brain, to the reproductive cycle and other systems. They know that by withholding information, they are prolonging the suffering they began, compounding the injuries they originally caused.

What people like Bob and Joe did was mislead people and doctors to think that MIC is nothing but a potent tear gas, scuttle the use of the only antidote known (sodium thiosulphate), send spin doctors and Pentagon toxicologists as specialists to help the Bhopal people, financially ruin the Red Cross Hospitals that were running in Bhopal, and much more.

Bob I hear has retired. What has become of Joe I don't know. Bob's position has been renamed Corporate Communications Manager and is held by Tom Sprick. Mahesh Mathai, maker of the movie, "The Bhopal Express" invited Tom to the New York premiere in April this year. Tom declined on behalf of Union Carbide but assured Mahesh that "the tragedy continues to be a source of anguish for the company." Tom is just another guy with a normal career, he probably even sleeps well each night.

Editors' Note: Opinions included here are that of the author. Robert Berzok accepts that he disagrees with the author on many of the points relating to the Bhopal tragedy, but denies meeting Satinath Sarangi on the night in question above.

Source: Satinath Sarangi, March 2001, from www.indiatogether.org

An open letter to all Dow employees on the eighteenth anniversary of the Bhopal tragedy (2002)

In this letter to all Dow employees, then CEO of Dow, Michael Parker, presents the company's position that it will never accept responsibility for what its wholly owned subsidiary did to the people of Bhopal.

(Note: This letter was posted for all Dow Employees on Dow's Intranet.)

Midland, MI - November 28, 2002

Dear Colleagues,

3 December marks the eighteenth anniversary of the terrible tragedy that occurred in Bhopal, India. It was a tragedy of unprecedented proportions, and no one in industry – especially the chemical industry – should ever forget. Indeed, as I have said before, I can still recall the exact moment I heard the news, and the profound sadness I felt. On 3 December, I plan to spend a few quiet moments reflecting on the lessons of Bhopal. As I do so, I will also personally recommit to achieving excellence in Dow's environment, health and safety performance, and continuing our drive toward Sustainable Development. I encourage every Dow employee to do the same.

In case you are unfamiliar with what took place in Bhopal 18 years ago, here's a very brief summary:

- Shortly after midnight on 3 December 1984, methyl isocyanate gas (MIC) leaked from a storage tank sited at a pesticide manufacturing facility in Bhopal.
- As it leaked from the tank, the gas drifted across the neighboring communities with devastating consequences. According to the Indian government, some 3,800 people died and thousands more were injured as a direct result of exposure to the lethal fumes.

Without a doubt, the tragedy changed our industry forever as companies across the globe collectively took on the moral responsibility to prevent anything like it from ever happening again. Indeed, the horrific event in Bhopal was the driving force for the design and implementation of Responsible Care®.

At the time of the disaster, the Bhopal plant was operated by Union Carbide India Limited (UCIL), a 51 percent affiliate of Union Carbide (Indian government financial institutions owned 26 percent of the shares and some 24,000 private Indian citizens owned the balance.)

As you know, Dow acquired Union Carbide's stock on February 6, 2001. And that is why, with the anniversary of the Bhopal tragedy approaching, it is very likely that Dow will be the focus of protests and demonstrations. In particular, we expect the environmental group Greenpeace to intensify their public campaign against us, strengthening their demand that Dow take responsibility for the tragedy. To provide some balance to the claims you'll likely hear from Greenpeace over the coming weeks, I wanted to reiterate Dow's perspective on this issue.

In the eyes of the highest courts of India, the Bhopal case is closed. In 1989 a settlement agreement was reached between Union Carbide, Union Carbide India Limited and the Indian government through which Union Carbide paid $470 million in compensation, covering all claims relating to the incident. In response to public concerns, the Indian Supreme Court reviewed this settlement agreement and, in 1991, determined it should stand – concluding that it was "just, equitable and reasonable."

At that point the legal case was closed. So when Dow completed its acquisition of Union Carbide stock in February 2001, the subsidiary had no remaining liability for the tragedy that had occurred 16 years previously.

The black-and-white legal case is one thing. But, there is more for us to consider. As a company committed to Sustainable Development, and therefore, the very notion of good corporate citizenship, we also have an entirely separate humanitarian question with which we have to wrestle. Since acquiring Union Carbide, we have been engaged in thoughtful discussions to answer the question – Can, and should Dow, in its role as a global corporate citizen, help to address any of the present day needs which are apparent in Bhopal? That is why, despite the fact that we clearly have no legal obligations in relation to the tragedy, we have, for some time, been exploring various philanthropic initiatives which might address some of those needs – just as we do in other parts of the world where we have business interests.

As a result, for the past two years we have held a number of discussions with the International Campaign For Justice in Bhopal in India, trying to better understand their concerns and to solicit help in identifying appropriate humanitarian aid projects. This work continues – and we remain hopeful that we can find an appropriate initiative in the not too distant future.

But what we cannot and will not do – no matter where Greenpeace takes their protests and how much they seek to undermine Dow's reputation with the general public – is accept responsibility for the Bhopal accident.

It is therefore extremely likely we will face a number of protests at our sites around the world during the coming weeks, and into the future. I regret for this distraction – I realize it can be both disruptive and distressing – but I hope you can understand why we will not yield to this sort of pressure.

I also hope you will not let this deter your pride in our company and all that it stands for. The products we produce benefit people around the world, improving their lives each and every day. Our commitment is clear. From our far-reaching and voluntary Environment, Health and Safety Goals for 2005 to our 12-point Implementation Plan for Sustainable Development, we at Dow will continue to strive to achieve our vision of zero harm to the environment, to our people or to anyone we touch in the value chain. So, on December 3rd, take a moment, to reflect on the tragedy of Bhopal, and to recommit to doing your part to moving our company ever closer to that vision of zero.

Thank you for your continued support.

Sincerely,

Michael D. Parker
President and CEO
The Dow Chemical Company

Source: www.dow.com

Public Relations Services (2004)

The Dow Corporation contracted Brown, Nelson & Associates to manage its website on Bhopal, www.bhopal.com. Excerpted here is the description of the capabilities of this Houston, Texas firm in its own words.

- Brown, Nelson & Associates, Inc.
- Brown Nelson specializes in PR for petrochemical, maritime, business-to-business, energy, and professional services.
- Clients most often request these public relations services:
- Issues & Risk Management
- Employee Communications
- Research
- Community Relations
- Management Counseling
- Media Relations

THE BROWN NELSON CRISIS RESPONSE MODEL

Today, the Brown Nelson model for public relations response is considered the industry standard. The agency has successfully tailored it for a variety of industries. The basis of the approach is to manage the public affairs aspects of an incident, not simply react to it. There are five major components...

1. Preparing and maintaining a public communications plan responsive to any company needs…
2. Maintaining a response team of crisis-experienced local public relations counselors for immediate reaction to any incident.
3. Operating simultaneous news centers on-scene and in Houston.
4. Providing senior-level public relations counsel to the company's incident team leader and other key members of that team.
5. Periodic tests and drills of the client's response plans and procedures.

Source: Brown Nelson & Associates in Houston, Texas.
Reprinted from www.bhopal.com.

Letter exchange between Dow/Carbide and the Bhopal Memory Project (2004-2005)

In Spring 2004, the Bard Human Rights Project responded to an information request about their Bhopal project from the email address horsenuzzler@aol.com. Immediately after referring "HN" to an unpublished draft of the project, HRP was contacted by Dow and then by UCC in an uncharacteristically long exchange of letters about Bhopal.

To: Bard Human Rights Project hrp@bard.edu
From: jmusser@dow.com
Date: 9 August 2004

I would appreciate an opportunity to speak to someone regarding inaccurate statements made on your "Bhopal Memory Project" web site. I recognize there are many differences of opinion regarding the Bhopal tragedy. I'm happy to share our opinion on these matters but I'm most interested in setting the record straight with you as to matters that are factually incorrect as posted on your web site. Thank you for your consideration.

Sincerely,
John Musser
The Dow Chemical Company
47 Building, Midland, MI 48667
989/636-5663 (O) 989/638-7800 (OFAX)
jmusser@dow.com

From: jmusser@dow.com
Subject: RE: Bhopal Memory Project
Date: November 11, 2004 12:00:12 AM GMT+05:30
To: bhanna@bard.edu

Dear Ms. Hanna,

Thank you for responding. I appreciate your consideration of the material I forwarded and I'm happy to review the site in its entirety and provide you with feedback. I hope to be back to you by the end of this week or early next.

Best wishes,

John Musser
Tomm F. Sprick
Director
Union Carbide Information Center

Nov. 18, 2004
Ms. Bridget Hanna
Bard College
Annandale-on-Hudson, NY
Bhopal Memory Project
bhanna@bard.edu

Dear Ms. Hanna,

John Musser at Dow has asked me to provide comments on the Bhopal Memory Project since Dow has no responsibility for Bhopal, and since I am Union Carbide's spokesperson regarding the tragedy, especially pertaining to all the relief and remediation efforts and the settlement.

I have completed a review of most of your proposed website content, but was unable to access all of the documents. That not withstanding, I offer the following comments on various sections:

In the "*About the 1984 disaster*" (and other) section(s) you say: a) "...at a neglected Union Carbide pesticide factory..." That is not an accurate statement.

First, the plant in Bhopal was managed, run and operated on a day-to-day basis by Union Carbide India Ltd. (UCIL) – not Union Carbide (UC). They two are separate companies in the eyes of the law, in the U.S. and in India. Union Carbide India Limited was a publicly traded company, with stock held by Union Carbide, Indian financial institutions and private investors. This is important because the Indian government essentially forbid Union Carbide to have any hands-on involvement in managing or operating the plant in Bhopal. It is also important to understand that UCIL was a very significant company, with more than a dozen facilities in India at the time of the tragedy (and still is, though the name of the company was changed to Eveready Industries India Ltd).

Second, little evidence exists to call the factory neglected. In fact, in 1982 a technical audit team from Union Carbide visited the plant to conduct a routine process safety review (a standard practice at all UC subsidiaries). All safety issues identified during that audit were addressed by the plant well before the December 1984 gas leak and none had anything to do with the incident.

b) Refer to the safety systems not being functional at the time of the tragedy in anticipation of the pending sale of the factory.

First, the facts show – based on several investigations – that the safety systems in place could not have prevented a chemical reaction of this magnitude from causing a leak. In designing the plant's safety systems, a chemical reaction of this magnitude was not factored in for two reasons: the tank's gas storage system was designed to automatically prevent such a large amount of water from being *inadvertently* introduced into the system; and the process safety systems were in place and operational and would have prevented water from entering the tank by accident. The system design did not, however, account for the *deliberate* introduction of a large volume of water.

Second, during a global meeting of UC Agricultural Products in mid-1984 (months before the disaster), managers had a "think piece" that included an idea about moving the business from India, but it was never seriously considered or acted upon. A document containing this information is one of the oldest included in the initial court case.

c) State that Union Carbide has at times "implied" the tragedy was caused by sabotage. We have always emphatically stated that the tragedy was caused by sabotage. You make no reference to the basis for our assertion — the Arthur D. Little investigation. Since you do offer the report as a reference at a different location on the site, I think it would be appropriate to note our basis and include a link to the report in this opening section of your site.

d) Assert that taking "moral responsibility…was a meaningless fig leaf…." Our then chairman, Warren Anderson, took his "moral responsibility" quite seriously and backed it up with immediate offers of monetary, medical and technical aid. For his effort of flying immediately over to India, against the advice of counsel, he was rewarded by being arrested, placed under house arrest and ultimately told to leave the country. He also saw all early UC efforts to provide relief rejected by the Indian government.

e) State that "no party has ever been charged or publicly identified...." That is only partially correct. Union Carbide shared the details of its investigation with the Indian government and the Indian authorities are well aware of the identity of the employee and the nature of the evidence against him. While he has never been charged is a question best answered by the Indian government.

In the *"20 Years Later"* section you:

a) Say that "hazardous materials were removed by UCC." Union Carbide never remediated the site. Up and through 1994, UCIL spent more than $2 million on site cleanup, with all the work being conducted under the strict oversight of the Indian Government. After Union Carbide sold its shares in UCIL and the company was renamed Eveready Industries Limited, we understand EIIL continued the remediation effort, again under strict oversight of the Indian Government. Then, in 1998, the government of Madhya Pradesh took control of the site and publicly assumed all responsibility for future remediation activities.

b) Reference "tests conducted by Union Carbide Corporation...noting that all groundwater samples caused immediate 100% mortality in fish." While I have no knowledge of such tests, you casually omit any reference to UC's publicly stated position that, "the government tested it and found nothing wrong with it." I know you were advised earlier about a report issued by the India's highly respected National Environmental Engineering Research Institute (NEERI), which, in 1997, found soil contamination within the factory premises at three major areas that had been used as chemical disposal and treatment areas. However, the study also found no evidence of groundwater contamination outside the plant and concluded that local water-wells were not affected by plant disposal activities.

Furthermore, a 1998 study of drinking-water sources near the plant site by the Madhya Pradesh Pollution Control Board did find some contamination, but did not find any traces of chemicals linked to chemicals formerly used at the UCIL plant. Rather, the Control Board found that the contamination likely was caused by improper drainage of water and other sources of environmental pollution. The point is, these reports were not made up; they do exist and that is what they say. A failure to mention these is a disservice to those objectively interested in all the facts.

While we have seen conflicting reports currently being made by various groups and media, we have no firsthand knowledge of what chemicals, if any, may remain at the site and what impact, if any, they may be having on area groundwater. We believe it is important for the State of Madhya Pradesh to restart and complete the remediation of the plant site. The state is in best position to evaluate all available scientific information and to make the right decision for Bhopal. For specific details, you'll need to contact the government of Madhya Pradesh.

c) Say "...but Union Carbide says that it was a saboteur...." While you may not believe Union Carbide's position, we do have the Arthur D. Little study to back our position. To my knowledge, there has been nothing but anecdotal evidence to support other theories.

Refer to the criminal indictments of individuals responsible for the disaster. All of the key people from UCIL – officers in the company and those actually running the plant – have appeared to face charges, which were reduced to a misdemeanor status.

Union Carbide officials have not been in court to face charges. They are not subject to the jurisdiction of the Indian court since they did not have any involvement in the operation of the plant. Therefore, it would be totally unfair to bring criminal charges against them.

In the about *"The Memory Project,"* section, you:

a) Again refer to the Bhopal plant as the Union Carbide plant. The correct reference would be "the Union Carbide India Ltd. or UCIL plant."

b) Say "Indeed, and had not Union Carbide spent millions of dollars on public relations, misinformation and evasion, perhaps the lives of the affected people would be different – and perhaps more of them would be alive today to tell about it." This statement is grossly unfair and very reminiscent of typical activist rhetoric that also fails to account for what Union Carbide did to provide assistance. No where on your site do I see any mention of the following:

- In the wake of the release, Union Carbide provided immediate and continuing aid to the victims and set up a process to resolve their claims.
- In the days, months and years following the disaster, Union Carbide took the following actions to provide continuing aid:

1) Immediately provided approximately $2 million in aid to the Prime Minister's Relief Fund.
2) Immediately and continuously provided medical equipment and supplies.
3) Sent an international team of medical experts to Bhopal to provide expertise and assistance.
4) Funded the attendance by Indian medical experts at special meetings on research and treatment for victims.
5) Provided $5 million to the Indian Red Cross relief fund.
6) Provided a $2.2 million grant to Arizona State University to establish a vocational-technical center in Bhopal, which was constructed and opened, but was later closed and leveled by the government.
7) Offered an initial $10 million to build a hospital in Bhopal; the offer was declined.
8) Provided an additional $4.6 million in humanitarian aid to victims.
9) Established an independent charitable trust for a Bhopal hospital and provided initial funding of approximately $20 million. Upon the sale of its interest in UCIL, and pursuant to a court order, provided about $90 million to the charitable trust for the hospital.

- Then, of course, there is $470 million settlement paid by UC and UCIL.

Together, all of the above should count for something more than "public relations."

My final concern relates to the [student- eds.] paper authored by [name omitted-eds.] The paper is riddled with factual errors and unsubstantiated opinions. For example:

The Bhopal factory was not closed in 1980.

I have already addressed the premise, that of the factory not being well maintained.

All the deficiencies identified in the 1982 audit were addressed by UCIL prior to the tragedy and did not contribute to it.

The plant's safety systems did not fail; they were not designed to be capable of handling a reaction of this magnitude.

Union Carbide does not now have, nor has it ever had, a figure for the number who died or were injured. All figures we discuss come from the

Government of India. And, perhaps most telling, are the comments she makes under the section "The Legal Aspect" in which she notes a discrepancy on the UC website regarding the dollars contributed by UCC employees to the relief efforts. Her comment – "this does not speak highly of the company's organizational and communicational abilities" – is quite ironic given that on page 1 of her own abstract she says that UC owned 59 percent of the stock in UCIL, while on page 13 she says the amount is 51.9 percent. And, the sale of the stock took place in 1994 – not 1992.

I appreciate your interest in hearing our perspective of your website. However, its tone, content and lack of content representing the "other side of the story" suggests that your objective is more that of advocacy, rather than one of providing a balanced record of the Bhopal tragedy and its aftermath.

Thank you again for seeking our input.

Sincerely,
Tomm F. Sprick
Director
Union Carbide Information Center

From: bhanna@bard.edu
RE: bhopal memory project
Date: December 20, 2004 5:31:46 PM GMT+05:30
To: mediarelations@unioncarbide.com

Tomm F. Sprick
Union Carbide Corporation
mediarelations@unioncarbide.com

Dear Mr. Sprick,

Thank you for your response to my genuine inquiry as to the positions of the Dow corporation on the content of my website.

You wrote that the website's "tone, content and lack of content representing the 'other side of the story' suggests that [my] objective is more that of advocacy, rather than one of providing a balanced record of the Bhopal tragedy and it's aftermath." Although much of the content has at this point been verified to my satisfaction, I remain open-minded, and committed to one of the most important functions of the website – that of creating an undiscriminating archive. I am therefore more than willing to address the mentioned "lack of content." Any further documentation or statements that you can provide me with will be posted as soon as possible. In fact, I would provide a whole page on my site devoted to your material.

I especially appreciated your clarification on several factual points; in particular about the types of remediation and support that Union Carbide has provided to Bhopal since the disaster. On other issues, I still have a few further questions about your position…

Sincerely,
Bridget Hanna
Bhopal Memory Project

From: mediarelations@unioncarbide.com
Subject: RE: bhopal memory project
Date: December 30, 2004 9:42:42 PM GMT+05:30
To: bhanna@bard.edu

Ms. Bridget Hanna
Bard College
Annandale-on-Hudson, NY
Bhopal Memory Project
bhanna@bard.edu

Thank you, Ms. Hanna…

For the opportunity to provide you with additional details concerning Union Carbide's position on the Bhopal tragedy. My company and I do, and I'm sure your readers will, appreciate the balance you provide to this issue.

Since 1984, the interest of journalists, citizens and students has risen and fallen many times as details surrounding the accident have were explained, have been forgotten, and resurfaced again and again. Literally, there has been no "new news" on the accident itself for many years. But, unfortunately for the Bhopal victims and their families, their problems have continued since the money from the settlement has languished in banks and promises to provide rehabilitation, medical assistance, vocational training and clean up of the site have not been kept by various government agencies. While today's journalists may feel they have "uncovered" new details, all of these information was examined and properly explained long ago.

I must add, however, that my comments in no way diminish our feelings of compassion for the victims and their families, and our feelings that the remaining funds should be distributed immediately to them and concerns raised about the environmental conditions at and surrounding the site addressed by the government. The government assumed ownership of the site in 1998 and, at that time, said it would clean up the contamination. Nor do I question the sincerity of the efforts of the victim's groups and environmentalists and/or other concerned organizations or individuals who continue to press the government to help the victims and their families.

Getting back to the accident itself, I'm sure you can understand – having put together the information for your website – how much has been written about the tragedy in the past 20 years. Faced with continuing worldwide interest, especially during five-year anniversary periods, Union Carbide long ago decided that we should concentrate solely on the key issues; that is, the plight of the victims and their continued suffering, rather than the minutiae of the accident itself, which, as I have said, has been reviewed to exhaustion in the last 20 years.

However, since you have devoted so much time and effort to your project and are interested in presenting a balanced view, let me try to address the additional questions you have raised:

1. Can you please provide documentation of both the safety review and the confirmation that "all of the issues ...were addressed by the plant"? If that is true then the plant certainly does not qualify as neglected.

A. I do not have, nor have I ever had a copy of the detailed operational safety survey to which you refer. I will do my best to try to obtain a copy for you, but cannot promise it.

What I do have, and am attaching to this e-mail, is a copy of a press release that Union Carbide issued on Dec. 11, 1984. When released to the media, this press release included a copy of the detailed operational safety survey for Union Carbide India Limited. As you'll read in the last paragraph of the press release, "...Subsequent progress reports from Union Carbide India Limited represented to Union Carbide Corporation that all problems uncovered in the 1982 survey had been take care of, except for...."

The last sentence notes: "We (UC) have no reason to believe that what was represented to us by Union Carbide India Limited did not in fact occur."

Bridget, it is important to remember that none of the deficiencies identified in the 1982 survey had anything to do with the cause of the tragedy as we know it.

2. I did not see a question No. 2

3. Re: 1e. Did you provide an actual name of a saboteur to the Indian government to assist in their search for the culprit of the crime?

A. Union Carbide never publicly disclosed the name of the employee because it would serve no useful purpose; UC is not a governmental body and has no authority to "arrest" or "charge" anyone. However, the Indian Government, through its Central Bureau of Investigation (CBI), had access to the same information as Union Carbide did. The CBI was well aware of the identity of the employee and the nature of the evidence against him. Indian authorities refused to pursue this individual because they, as litigants, were not interested in proving that anyone other than Union Carbide was to blame for the tragedy. The fact that employee sabotage caused the disaster under existing law would have exculpated Union Carbide. You may be interested to note that the CBI subjected the UCIL employee who found the local pressure indicator was missing on the morning after the accident (a key factor in UC's sabotage theory) to six days on interrogation to get him to change his story. That effort was unsuccessful.

4. Re: 2b. The document including the quote about the 100% mortality in fish is attached. However, I don't have any independent source to verify that this is an official Carbide document. Can you help me with that? Is it a Carbide document? I will make sure to mention the NEERI report on the site as well, but I haven't seen the Madhya Pradesh Pollution ControlBoard study. Can you please send it to me? A. I did not see any attachment re: the fish study. If you want to resend it, I will look at it, but can't promise that I will be able to verify it.

Separately, as you requested, I have attached a copy of the MP Pollution Control Board press release from July 1998 that references the 230-page 1997 NEERI study. I started to underscore several of the key conclusions the MP government makes for your ease of reference, but then realized the entire document is worthy of your review and quite telling.

In case you have not heard, you should be aware that a BBC news story on Dec. 2, 2004, and confirmed Dec. 3 from India quotes Uma Shankar Gupta, minister in charge of the [Bhopal] gas relief department, as saying that the Indian federal government has asked an Indian company to carry out a survey of the site to assess the extent of the problem [contamination]. Additionally, a Dec. 13 Press Trust of India story quotes another government official as saying Engineers India Ltd has submitted a technical proposal for removing toxic wastes lying in and around the former plant, and that the plan was under consideration. If these latest stories are accurate, Union Carbide is encouraged and hopes that the Indian government initiates the clean up as soon as possible. The state is in the position to evaluate all scientific information that is available and make the right decision for Bhopal.

5. Re: 3b. You wrote "in the wake of the release, Union Carbide provided immediate and continuing aid to the victims." What kind of aid, and how was it distributed? If that is true, it makes a huge difference in this issue. Please send me the details.

A. Union Carbide immediately accepted moral responsibility for the tragedy, and this commitment was reflected in our relief efforts, which included:

1) Contributing $2 million to the Indian Prime Minister's relief fund. This money initially was rejected by the government, but later was accepted.
2) Donating $5 million to the Indian Red Cross's relief fund. This money initially was offered to the Indian Government, which refused it. Then UC contributed it to the Indian Red Cross.

Please note that in both of these cases, as was true with the final settlement monies, Union Carbide had/has no way of knowing how the monies were used, who benefited and how it may have ultimately been distributed.

3) Organizing and sending to India a team of top medical experts to help identify the best treatment options and work with the local medical community.

4) Marshalling all the information on the toxicity on the gas methyl isocyanate (MIC) and providing substantial expertise for the short-term and long-term treatment of victims. On the day of the tragedy, UC also dispatched a team of technical MIC experts, who carried all published and unpublished studies on MIC available at that time and shared those studies with medical and scientific personnel in Bhopal.

5) Providing substantial amounts of medical equipment, supplies and expertise to the victims.

6) In the first half of 1985, UC's U.S. employees, retirees and former employees collected and distributed $120,000 to Bhopal relief organizations.

7) Granting, in April 1985 and January 1986, $2.2 million to Arizona State University to set up and operate a vocational-technical center in Bhopal to provide training and local jobs. Please note that the Government of India demolished this center in 1987 after discovering that Union Carbide had funded its development and construction through the aforementioned

grant. I arrived in Bhopal one day after the demolition took place and saw the destruction!

8) Funding, in June 1985, visits to the U.S. by Indian medical experts to attend special meetings on research and treatment of victims exposed to MIC.

9) Donating, in May 1986, $1 million to Sentilles, a Swiss-based humanitarian organization, for medical, educational and training programs in Bhopal.

10) UC's efforts culminated in 1989 with the settlement and, later, the contribution of $100 million from our sale of the UCIL shares to a charitable trust fund to build a hospital for victims. The hospital became operational in 2000.

6. What was the "additional $4.6 million in humanitarian aid to victims" spent on, and who administered it?

A. During the mid 1980s, UC "consolidated" – for ease of reference – some of its relief dollars into one figure, rather than detail them as I have for you in this document. Some journalists prefer the details; other want "ballpark figures." The $4.6 million figure used early on was actually a bit more than $5 million and represented Nos. 2 and 6 in the answer to question No. 5 above. I have recently updated our website to make this consistent. Again, UC does not know how the monies were used, who benefited and how it may have ultimately been distributed.

I hope you find this additional information of interest and will be incorporated into your project's website. Again, my thanks and have a Happy New Year.

Best regards,

Tomm F. Sprick
Director
Union Carbide Information Center
mediarelations@unioncarbide.com

12

Responsibility

Can Bhopal make the corporation accountable?

BHOPAL FOREVER ALTERED the image of both the chemical industry and of chemicals themselves. After Bhopal, the imagery of the green revolution and of the romance of chemistry disappeared because they were no longer convincing. With demands internationally, particularly in the United States, for freedom of information and more stringent regulations immediately after Bhopal, the chemical industry had to act quickly to create a new image – one that emphasized the industry as a protector of the environment rather than its destroyer and as an actor that tried to preempt regulation by creating a system of apparent self-regulation. Called *Responsible Care*, this initiative and its attendant imagery – dubbed "greenwash" by many outside the industry (see the first excerpt below) – have dominated the industry ever since.

Although the corporate damage control since 1984 has been largely successful – neither Union Carbide nor Dow has yet to make any major concessions (although the anniversary "Hoax," detailed below, imagines that very scenario) – it has continued to be just that and no more. Meanwhile, the fronts that the industry is defending have continued to grow – because of the increasingly commonplace nature of environmental damage by industry, the longevity of chemical contamination, the tenacity of survivors and their oft-affected children, and the altered public perception of industrial danger. And, along with the merger of these two chemical giants –

resulting in the largest chemical company in the world with operations in 168 countries – came the incidental merger of their opponents' and victims' interests. Dow has a dossier of environmental devastation and human rights violations nearly as colorful as Carbide's, and the combined record is damning (Dow's record is covered in this chapter by Jack Doyle in *Toxic Trespass*). Whether they are victims of Agent Orange, Bhopal, or contamination in Michigan, the voices of these different groups are combining with one demand – make the corporation liable.

Simultaneously, shareholder activism – a relatively new front for political action – is beginning to have a serious impact. Last year, the holders of 6 percent of Dow's shares voted for the corporation to take responsibility for Bhopal. This came on the heels of a report (excerpted below) from Innovest, an investment firm, that condemned Dow as a poor investment based on their many continuing liabilities, including Bhopal. If further liability or criminal punishment for the twenty-year-old Bhopal disaster or the continuing water contamination is applied in the next few years, it will set an important precedent for corporate accountability worldwide. It could signify the beginning of the end of relative standards of safety and justice for corporations, particularly across international borders.

This chapter concludes with the hoax that convinced the BBC and the world, on the twentieth anniversary of the disaster, that Dow was taking full responsibility for Bhopal to the tune of $12 billion. (This hoax bore the fruit of what has been a pitched battle on-line for the opportunity of providing these different versions of the Bhopal story. UCC maintains www.bhopal.com, which they pay to advertise on Google. This website has no pictures; it contains only material that supports the position that the corporation is not liable – i.e. ADL, NEERI, and Carbide PR. Easily confusible, www.bhopal.net is the website of the International Campaign for Justice in Bhopal [ICJB] and www.bhopal.org is the site of the Bhopal Medical Appeal. The hoax concerns www.dow-chemical.com, now www.dowethics.com, the Yes-men's spoof on www.dow.com (see Claudia Deutsch's explanation below.) For the short hour before the hoax was exposed, this admission of responsibility was amazing to survivors, activists and the public, but it was no longer unthinkable.

Bhopal continues to be on the forefront of corporate activism, and, however incredible it might seem, a scenario like the one proposed in the hoax is beginning to appear as not only possible, but a likely, eventuality of this twenty year disaster.

The greenwashing of corporate culture (1996)

While community organizations around the world were demanding better regulation of chemical companies, the corporations took an ingenious step called "responsible care" – a self-regulation/public relations initiative for environmental safety.

As part of the greenwashing counterstrategy, corporations have notified the public that there has been a profound change in corporate culture. Some common manifestations of this new concern for environmental image and performance are:

- Corporate restructuring to include environmental issues, e.g., environmental officers at high levels, or new environmental departments within a corporation.
- Corporate environmental programs like waste minimization, waste reduction, and product stewardship.
- Responses to public concern about the environment; sometimes, these responses take place even when not required by law.
- Environmental themes in advertising and public relations.
- Voluntary environmental policies, codes of conduct, and guiding principles.

With the creation of such programs, we are asked to believe that corporations are now something fundamentally different than what they were before. But the addition of an "environmental department" does not change the *raison d'etre* of a corporation. It is critical that citizen activists and governments look under the surface of such announcements and be aware of the overall context in which they exist.

Certain basic characteristics of corporate culture have not changed. For example, in overseas operations especially, the assertion is often made that the mere presence of the corporation, its products, technologies, jobs, and culture are inherently beneficial to the host population. As an international waste trader, Arnold Andreas Kuenzler, said about a planned hazardous waste landfill in Angola, "If it's good enough for the Swiss, it's good enough for the blacks." In many cases, corporations further imply that the dirty industries they bring will be the primary method whereby Southern countries can gain enough wealth to have the "luxury" of a clean environment.

Another fundamental dynamic is that through aggressive marketing or interlocking relationships with "customers," manufacturers help create "demand." Corporations then proceed to abdicate responsibility for problems created by their products by passing responsibility to the users of those products. Responsibility is passed along by intermediate users (such as the automobile industry in the case of CFCs) until it reaches the individual consumer. In the end, individuals are held responsible for production and marketing decisions made by giant corporations. This has even been the excuse for marketing products in the South which have been banned or restricted in the North such as lead gasoline additive and some pesticides.

To understand why corporate culture is so impervious to change, it is necessary to bear in mind the brutally competitive global economic atmosphere – between countries as well as companies – which TNCs are both responding to and fueling. Richard. J. Barnet and John Cavanagh make the obvious but essential point that corporations are not chartered to solve social or environmental ills, but "are in business to make products and sell services anywhere they can to make money."

In their advertisements and slogans, however, TNCs are increasingly trying to give the impression that they are in business to help people and to solve environmental problems. DuPont's "Better Things For Better Living," and "Dow Lets You Do Great Things," are examples. People watching television or reading magazines with Greenwash may be seduced into forgetting that the fundamental drive of corporate culture is not improving their lives. Rather, the emphasis is on achieving cost advantage: minimizing expenses and maximizing revenue. In the culture in which business executives operate, the profit motive and pressure to raise shareholder returns remain the most influential determinants of corporate behavior and the key criteria for judging corporate performance.

Thus, despite their stated commitment to environmentalism, TNCs typically continue to justify their current activities, and new investments, with a cost-benefit assumption which fails to include the vast majority of environmental costs. Measuring only direct costs and short-term profits, corporations may tout the benefits of jobs and products created, but externalize costs of pollution, waste, and long-term damage to people and the environment.

Inside the World of Greenwash: Corporate Codes of Conduct

Many TNCs have adopted "corporate codes of conduct," "guiding principles," and other voluntary environmental policies as part of their response to ecological problems. Because these codes are offered as evidence that industry is taking its responsibilities seriously and is prepared to respond to citizen demands, they must be examined closely. Some companies imply that voluntary adherence to a code can replace regulations and monitoring of industry. Rhone-Poulenc has put this belief into practice. Its Chairman Jean-Rene Fourtou, who formed a 25-company association called "Enterprises pour l'Environnement," credits the group with success in heading off new regulations in favor of voluntarist initiatives.

Skeptical observers of the codes typically say that while the rhetoric is pretty, practice hasn't yet changed enough. But a closer look at the actual texts of these codes reveal that even the skeptics are too trusting. The codes adopt environmental terminology, such as "environmentally sound" and "sustainable development," while subtly changing the meaning of key words to cover industry behavior. In the end, the new rhetoric and the acknowledgement of relatively superficial problems in voluntary codes divert attention from the fundamental environmental issue: products such as nuclear reactors and toxic chemicals form the lifeblood of many TNCs. The codes are themselves a form of greenwash.

Two of the major corporate codes are the chemical industry's Responsible Care Program and the International Chamber of Commerce Business Charter for Sustainable Development, also known as the Rotterdam Charter.

Responsible Care

"Responsible Care" is the name of the chemical industry's major program on environmental issues. It originated in Canada in 1984 and was adopted in the U.S. as a direct response to the Valdez Principles, a code developed by environmental organizations for corporations following the 1989 Exxon Valdez oil spill. All members of the U.S. Chemical Manufacturers Association must sign on to Responsible Care as an obligation of membership. Chemical industry associations in western Europe and, more recently, Latin America, are developing similar programs, and it is a point of pride among many chemical company executives.

The two aspects of Responsible Care consistently emphasized by the associations and individual members are a "commitment to continuous improvement" in health, safety, and the environment, and the "profound cultural change" it represents. Responsible Care acknowledges that the chemical industry as a whole has not performed even to its own satisfaction and that change is needed. This gives citizens concerned about company practices some leverage, and is a welcome admission. But there are a number of serious problems with Responsible Care:

– The U.S Chemical Manufacturer's Association's president has stated that Responsible Care will help citizens to track corporations, monitor their performance, and make suggestions. Toward that end, each company is supposed to conduct an annual self-evaluation. However, the evaluations are not available to the public. Without access to information – even that generated by the company itself – the public does NOT have the opportunity to track the corporation any more than it did before Responsible Care.

- Although one of the "Guiding Principles" of Responsible Care is to develop Safe products, there are no criteria for what constitutes a safe product. Even the most dangerous products, such as banned pesticides and ozone-destroying chemicals, are judged "safe" by Responsible Care signers.
- Under the heading "Pollution Prevention Code," Responsible Care has two parts: waste and release reduction and waste management. While waste reduction is desirable, this blithe interpretation of "pollution prevention" makes the phrase meaningless. Waste reduction and management are often forms of end-of-pipe pollution control measures, not preventative measures. Pollution prevention should refer to the avoidance of toxic chemical production, use, and disposal in the first place. The text takes waste practices which are responsible for much of the pollution spread by TNCs and legitimizes them as "prevention."
- Responsible Care emphasizes "environmental performance," suggesting that the only thing wrong in the chemical industry is that there are too many "incidents." While a reduction in accidents and spills is vital, it is notable that Responsible Care does not acknowledge the inherent toxicity of many chemical company products and routine emissions. Thus, a corporation which increased production of an unnecessary and toxic product can claim to have improved "environmental performance" if it has had fewer accidents in the manufacturing process.
- Responsible Care does not apply to foreign subsidiaries of member companies. Company evaluations do not include overseas operations, and overseas environmental policies are not addressed. Other business charters and some companies are more comprehensive than Responsible Care in this area.

– Responsible Care has failed to take root in much of the chemical industry. In a 1992 survey of US Responsible Care member companies, the US Public Interest Research Group found that 42 per cent of the companies were unreachable by phone, and 27 per cent of those contacted refused or were unable to answer any questions about chemical use, storage, shipments, and related matters. Only ten per cent of the 192 facilities surveyed answered all the questions asked by the group. The Chemical Manufacturer's Association, which organized Responsible Care in the US, found in its own survey that relatively few chemical industry employees had heard of Responsible Care.

Source: From *Greenwash: The Reality Behind Corporate Environmentalism*, New York and Penang Malaysia, The Apex Press and the Third World Network, 1996. (References deleted)

Dow's trespass (2004)

A comprehensive examination of Dow's record by Jack Doyle concludes with this articulation of the company's long history of "toxic trespass."

Dow Chemical has been polluting property and poisoning people for nearly a century, locally and globally trespassing on workers, consumers, communities, and innocent bystanders; on wildlife and wild places; on the global biota and the global genome. Granted, these transgressions were not intentional – at least not initially, as young Herbert Dow pursued his craft in the 1890s. Yet, through the years, as the commercial and legal apparatus of the modern chemical industry evolved, as more was learned about toxic and chemical intrusion, and as businesses like Dow's embedded, rationalized, and defended their practice, the trespass became more intentional and knowing. For the trespass by Dow finds its basis in the harm of chemical molecules crossing into living cells; commercial creations that touch off a cancer, cause a mutation, alter hormonal messaging, or harm reproduction. Such consequence is owed, in part, to the intentional design of chemical molecules to do certain directed things for monetary return to – kill pests, to be durable, to dissolve grease and grime, and mostly, to last for a long time with those traits; i.e., to be persistently toxic. Knowledge of these commercially valuable and aggressive traits has been around for decades; and so has knowledge about their untoward effects on much of biology.

Lawyers, legislatures, and courts have labored long and hard to design institutional mechanisms for dealing with the explosion of synthetic chemicals in society and the wrongs they have caused.

Regulations abound, and all the parties involved seem comfortable with the process that has evolved – save victims and other unrepresented life. This regulatory system, supposedly in place to protect public health, is ponderous and slow; essentially a generation behind the carnage. And the burden of proof is wrong; placed on the victims instead of the perpetrators. Chemicals are innocent until proven guilty, even though most have never run a full gauntlet of safety tests. For the last 40 years or more, this process has played out in protracted chemical-by-chemical battles that typically favor "reasonable" phaseout time – tables that mostly give businesses more years of marketing time. Public and environmental health are marginally protected, at best. This is not a fair or reasonable system. A better solution is available now.

Dow Chemical, for starters, must end its toxic trespass by disengaging from the production of those chemicals, chemical products, and chemical processes that jeopardize public health. This means the POPs and the PBTs – persistent organic pollutants, and the persistent, bioaccumulative toxic substances. It also means those that cause cancer, birth defects; and/or genetic damage, and those that disrupt hormonal messaging, cause developmental changes, or affect intelligence. This obviously, is a tall order for any chemical company. It means nothing short of Dow revamping its corporate culture and strategic business plan; adopting a Hippocratic Oath styled outlook to do no harm; and not to let loose any chemical in the world until all of its toxicological effects have been thoroughly investigated and aired publicly. This will not likely happen voluntarily by Dow's hand, as Dow executives and attorneys will continue to defend their business and practice as they always have. Rather, changing Dow, and the chemical industry, will likely come from the outside.

Gathering Storm

There is a storm building around the Dow Chemical Company; a gathering storm of aggrieved, injured, and angry parties-those who are fighting for harms already done, as well as those tired of chemical intrusion, whether by workplace exposure, factory emissions, product leaching, or toxic legacy. In too many instances, "Dow brand" toxic material has been invasive, harmful, and life altering. The aggrieved and injured include Dow neighbors from Texas and Louisiana, and fourth generation Vietnamese children suffering Agent Orange's continuing harms. They are joining with labor, investors, lawyers, religious organizations, and everyday moms and dads to make common cause. They are aiming at Dow for specific and continuing harms, but also as corporate symbol and industry surrogate for a process that continues to invent and release synthetic chemicals that are dangerous and archaic. The battle with Dow will be joined on many fronts, by activists and everyday people, who will use whatever tools and avenues of appeal are available – legal, economic, or persuasive.

But on one front there will be more and better information; new knowledge about toxic "body burden" – the hundreds of chemicals seeping into people's blood and body tissue. Although not well understood today, the body burden concept is certain to change people's thinking about toxic chemicals. "In general the public is more aware of chemicals found in the fish they consume or the data from toxic release inventories than they know about the chemicals found in their own bodies," write public health activists Sharyle Patton and Gary Cohen. "But there is a deep psychological significance in knowing that the tissues of one's body are being used as a chemical storage site." Indeed, as body burden surveys become more commonplace, and more people realize that their personal space, and that of their children, has been violated by chemical manufacturers like Dow, the politics of public health in the toxic chemicals arena will likely escalate to a new and more powerful level.

And there is another dimension as well. Again, Sharyle Patton and Gary Cohen explain: "Internationally, several United Nations conventions support the human right to freedom from chemical contamination. [T]he United Nations Human Rights Commission has recognized the right to a non-polluted environment as a basic human right. The Convention on the Rights of a Child protect the child's right to integrity of person and right to the highest possible standard of mental and physical health. By anyone's definition of basic human rights, the fact that infants are starting life with a body burden of chemicals represents a gross violation of human rights."

Massive Experiment

One of the results of the Nuremberg Trials after World War II was a universal agreement that civilized nations should not engage in chemical experimentation on humans, even in times of war. "Yet for the last sixty years," argue Patton and Cohen, "the chemical industry has engaged in a massive chemical experiment on the world's human population and the entire web of life. No one has ever given their consent for this experiment. Most people don't even know it is happening."

Dow Chemical is a main player in this experiment; the world's largest manufacturer of some of the most troubling compounds now used in commerce. But Dow is not the kind of company that yields easily. Dow has already had a number of warning shots fired across its bow – from the 1960s with napalm and Agent Orange in Vietnam, through the 1970s with 2,4,5-T and other pesticides, the 1980s with its dioxin-harms cover-up, the 1990s with industrial pollution and the silicone fiasco, to the current battles over dioxin, asbestos, and Bhopal. After each crisis, Dow typically assures the public it has learned a lesson; a new corporate ethic is installed, a new vice president is named, and always, a pledge to do better. But despite the principles, pledges, and new vice presidents, the toxic revelations keep coming, the body burden grows, and the toxic trespass continues. The result, for the most part, is business as usual. But change is coming.

Source: Jack Doyle, *Trespass Against Us: Dow Chemical and the Toxic Century,* Monroe, Maine: Common Courage Press, 2004.

Dow Chemical: Risks for investors (2004)

Report prepared by Innovest Strategic Value Advisors, assessing Dow's prospects as a solid financial investment, given its continuing liability in Bhopal and elsewhere, illustrates that Bhopal may be beginning to penetrate the corporation where it matters – with the stockholders.

Key Issues for Strategic Investors:

• **Overall Market Risk:** Dow could be pressed by markets and regulations to reduce its production and marketing focus on organochlorine chemicals as well as many other chemicals in its product portfolio. Mounting scientific findings regarding organochlorine toxins, dioxins and furans in particular, could result in more momentum for widespread phaseout of the company's products such as vinyl chloride monomers (used in PVC plastics), as well as many of the company's pesticides. In addition, plasticizers such as phthalates have phaseout risk as well which could affect Dow's plastic-related business. As the practice of testing of human blood and other tissues for the presence of these substances grows, the result may be material levels of tort liability should the company be linked to negative health impacts caused by "persistent bioaccumulating toxins" (PBTs). Many of the company's products or pollutants associated with their manufacture, use and disposal are PBTs. Dow's business strategy appears to be fully committed to the further development of organochlorine chemicals and other chemicals with attendant PBT risk profile.

• **Agent Orange:** The recent Supreme Court decision (Stephenson v. Dow Chemical et al. June 9, 2003) may open the door for Vietnam veterans not covered under a previous settlement in 1984 to pursue compensation with Dow Chemical for health risks associated with the chemical defoliant commonly known as Agent Orange. Given the number of claims and the extent of damage alleged to be caused by Agent Orange, the proceedings could result in sizable ongoing liability. Numerous foreign veterans groups and Vietnamese citizens affected by Agent Orange exposure are also seeking compensation from manufacturers.

• **Bhopal:** The Bhopal disaster is an ongoing concern with significant potential to harm the company's reputation or pose material liabilities, as well as constrain investment in Asia. Continuing and heated controversy over reparations to victims, deaths and birth defects related to methyl isocyanate exposure, and pollution of the city's water supply could result in potential legal liability. Dow's wholly owned subsidiary, Union Carbide has been deemed an "absconder from justice" for failing to appear before the courts in India to face criminal charges stemming from the disaster. Efforts are underway in India to have the courts place responsibility on Dow to require Union Carbide to appear as a defendant in the criminal case. The $2.18 trillion market currently under SRI management worldwide may remove Dow as a potential investment as a result of these controversies. Dow management has flatly claimed that it has no liability associated with these matters, but our review indicates that it appears to have settled on an inadequate strategy to address the issue prior to merging with Union Carbide.

• **Contamination in Michigan:** Dow may incur potentially material liability related to dioxin contamination of more than 22 miles of the Tittabawassee River as well as sections of the Saginaw River and Saginaw Bay in Michigan. A class-action lawsuit involving more than 300 plaintiffs is currently in discovery. In the 2003 10K, Dow has reported an accrued $54 million in remedial liabilities for Midland – which appears inadequate to reflect the range of potential liabilities associated with this matter.

• **Semi-conductor Worker Liability:** Union Carbide, a wholly owned subsidiary of Dow since 2001, is currently involved in litigation stemming from the semiconductor industry, to which it is a supplier, involving claims of worker exposure to hazardous chemicals.

• **Current Financial Obligations:** The above issues, added to Dow's well known obligations under asbestos and breast implant litigation, and a $10.7 billion in debt and a debt-to-capital ratio of 53%, point to further strain on company reserves and thus increase the potential financial risks associated with Dow's overall product and environmental liability scenario.

Source: From a report prepared by Innovest Strategic Value Advisors, April 2004.

Bhopal critics in web hoax against Dow Chemical (2002)

Claudia H. Deutsch describes a 2002 stunt by The Yes-Men, an activist duo, who put up a website called www.dow-chemical.com, subtly parodying the actual Dow website and their position on Bhopal.

Last Tuesday, on the 18th anniversary of the lethal gas spill at a Union Carbide plant in Bhopal, India, that killed thousands of people, journalists received an e-mail press release claiming to be from Dow Chemical, which now owns Union Carbide. It was a fake, as was the Web site called up by a hyperlink in the e-mail.

The release supposedly explained why Dow refuses to clean up Bhopal or help people who remain sick from the spill. The link was to dow-chemical.com, a Web site that looked much like Dow's real Dow.com site, but that included such fake items as a "draft" of a speech by Dow's chief executive, Michael D. Parker, disavowing Dow's responsibility for Bhopal.

The hoax was the work of the Yes Men, a group of critics of business and government who gained attention in 2000 with Gatt.org, a bogus World Trade Organization site.

This time the Yes Men were too clever by at least half: they registered the site with Gandhi.com in the name of James Parker, Michael Parker's real son. So the younger Mr. Parker took ownership, and Dow took the site down last Wednesday night.

"We thought it would be funny, but it turned out to be stupid," said Andrew Bichlbaum, a Yes Men volunteer in Paris who set up the site. "We gave them the chance to claim the site as their property."

The Yes Men resurrected the site on Friday, as dow-chemical.va.com.au, whose host is Virtual Artists, an Australian company. Any visitor can download a copy of the site, Mr. Bichlbaum said, "so that if Dow gets this one too, it will continue to exist."

Although no other environmental group has acknowledged participation in the hoax, at least one voiced approval. "We support the people who published this site," said Casey Harrell, the Bhopal specialist at Greenpeace, one of Dow's most vocal critics.

Dow, meanwhile, maintains that the Web site violated numerous cyberspace copyright laws.

"It is ironic," said John Musser, a Dow spokesman, "that groups that position themselves as public defenders against companies that act irresponsibly, unethically or unlawfully are turning out to be the poster children for those very behaviors."

Source: From *The New York Times*, December 9, 2002

The BBC has fallen victim to an elaborate hoax (2004)

On the twentieth anniversary of the gas disaster, the BBC contacted the wrong website for comment from Dow Chemical. For a few minutes, the world took seriously the possibility of justice for Bhopal.

BBC World and BBC News 24 ran an interview with a bogus Dow Chemical official who claimed the company admitted responsibility for the Bhopal disaster in 1984.

He also claimed the company had established a $12 billion fund to compensate victims' families and survivors of the disaster.

Excerpts from the interview were also carried on news bulletins on Radio 2, Radio 4 and Radio Five Live.

The BBC has apologised to Dow and to viewers who may have been misled.

Elaborate deception

However, the BBC later admitted that the interview with bogus Dow spokesman Jude Finisterra was part of "an elaborate deception" and everything he said was false.

A statement read out on BBC World said: "This morning at 9 a.m. and 10 a.m. (GMT) BBC World ran an interview with someone purporting to be from the Dow chemical company about Bhopal.

"This interview was inaccurate, part of an elaborate deception.

"The person did not represent the company and we want to make it clear that the information he gave was entirely inaccurate."

A correction was also read out on BBC Radio news bulletins.

It said: "Earlier this morning, our news bulletin here (on Radio 2/4/5 Live) carried an extract from an interview with someone purporting to be from the Dow chemical company about the disaster twenty years ago at Bhopal in India.

"It is now clear that the person did not, in fact, represent the Dow company and we want to make clear that the information he gave was entirely inaccurate."

"No basis whatsoever"

Dow Chemical spokeswoman Marina Ashanin told BBC World from Switzerland: "Dow confirms there was no basis whatsoever for this report.

"We also confirm Jude Finisterra is neither an employee nor a spokesperson for Dow."

The BBC is looking into the incident to establish the background and how the interview got to air.

A report will be made to the BBC's Deputy Director-General, Mark Byford.

Thousands were killed instantly on December 3, 1984 when the Union Carbide plant in Bhopal released 40 tons of lethal methyl isocyanate gas into the air, in one of the world's worst environmental disasters.

Rachna Dhingra, a spokesperson for the International Campaign for Justice in Bhopal, said: "It is a cruel, cruel hoax to play on the people of Bhopal on the 20th anniversary of this tragedy."

Source: By Matt Holder, 3 December, 2004, BBC News Watch.

Dow's retraction of the settlement story (2004)

Dow reacts to the "unfortunate situation" caused by a hoax that announced that they were taking responsibility for the Bhopal disaster.

BBC Retracts Hoax Story about Bhopal

Midland, MI - December 03, 2004

Earlier today, the BBC was the subject of a hoax by an activist who falsely identified himself as a Dow employee. BBC has retracted the erroneous story.

Please visit the BBC Web site for their retraction.

Dow confirms that there was no basis whatsoever for this report. BBC has been informed of this error and has pulled the erroneous story. According to a statement issued by the BBC, "This information was inaccurate, part of an elaborate deception. The person did not represent the company. We want to make it clear the information he gave was entirely inaccurate."

This is an unfortunate situation for everyone involved.

The Bhopal tragedy occurred 20 years ago today. Links to statements from Union Carbide Corporation and The Dow Chemical Company on the anniversary are attached below.

Union Carbide's statement can be found at www.unioncarbide.com/bhopal.

Dow's statement can be found at www.dow.com/bhopal.

Source: www.dow.com

The Yes-Men tell the story of the hoax (2004)

Andrew Bichlbaum, who posed as "Jude Finesterra," Dow spokesman, and his partner in crime, tell the story of the hoax that shook Dow, the BBC and all of Bhopal.

On November 29, an email comes in to DowEthics.com: BBC World Television wants a Dow representative to discuss the company's position on the 1984 Bhopal tragedy on this, its 20th anniversary.

Knowing Dow's history of gross negligence on this matter, we think it unlikely they will send a representative themselves, and if they do, he or she will likely only reiterate the old nonsense yet again, which will be depressing for all concerned. Yes, we'd better just do their PR for them.

Since we can't possibly afford to go to London with our pathetic American dollars, we ask to be booked in a studio in Paris, where Andy is living. No problem. Mr. Jude (patron saint of the impossible) Finisterra (earth's end) becomes Dow's official spokesperson.

What to do? We briefly consider embodying the psychopathic monster that is Dow by explaining in frank terms how they (a) don't give a rat's ass about the people of Bhopal and (b) wouldn't do anything to help them even if they did. Which they don't. This would be familiar territory for Andy: he did something similar representing the WTO on CNBC's Marketwrap.

Instead we settle on having the impossible Jude announce a radical new direction for the company, one in which Dow takes full responsibility for the disaster. We will lay out a straightforward ethical path for Dow to follow to compensate the victims, remediate the site, and otherwise help make amends for the worst industrial disaster in history.

There are some risks to this approach. It could offer false hope to people who have endured 20 years of suffering because of Dow and Union Carbide. But what's an hour of false hope to 20 years of unrealized ones? If it works, this could focus a great deal of media attention on the issue, especially in the US, where the Bhopal anniversary has often gone completely unnoticed. Who knows, it could even somehow force Dow's hand.

After all, the real hoax is Dow's claim that they can't do anything to help. They have conned the world into thinking they can't end the crisis, when in fact it would be quite simple. What would it cost to clean up the Bhopal plant site, which continues to poison the water people drink, causing an estimated one death per day?

We decide to show how another world is possible, and hope that it's worth it....

Another problem we anticipate is that this could result in some backlash for the BBC. This is bothersome, because they have covered Bhopal very well, infinitely better than what we're used to in the US. We would much rather hoax CBS, ABC, NBC, CNN, or Fox, but none of those could give that rat's ass about Bhopal, and so none of those has approached us.

In any case, it didn't seem to hurt CNBC when "Granwyth Hulatberi" appeared as WTO spokesperson. It was a simple mistake, and one that anyone could make. Intelligent people will not question the excellence of BBC's overall coverage because of an unavoidable mistake, especially if it is caught quickly and provides for some interesting discussion that wouldn't have happened otherwise.

On the day of the interview, we wake up early and put on our thrift-store suits. Andy nervously runs through his answers once more while Mike fumbles with cameras. A crowded metro ride later, we arrive at the BBC's Paris studio. "Jude" is seated in front of a green screen and waits.

At 9 a.m. GMT, Dow's spokesperson appears live on the BBC World Service in front of the Eiffel Tower. He is ecstatic to make the announcement: Dow will accept full responsibility for the Bhopal disaster, and has a $12 billion dollar plan to compensate the victims and remediate the site.

They will also push for the extradition to India of Warren Anderson, former Union Carbide CEO, who fled India following his arrest 20 years ago on multiple homicide charges.

When it's over, the studio technician is happy about what she has heard. "What a nice thing to announce," she says. "I wouldn't work for Dow if I didn't believe in it," replies Andy matter-of-factly.

The full BBC interview runs twice, and the story quickly develops into the top item on news.google.com. It takes Dow two hours to notice that alas and alack, it's done the right thing: it must then emphatically state that it is not in fact doing anything of the sort. The retraction remains the top google story for the rest of the day.

Back at Andy's apartment, we help Dow express itself by mailing out a more formal retraction: "Dow will NOT commit ANY funds to compensate and treat 120,000 Bhopal residents who require lifelong care.... Dow will NOT remediate (clean up) the Bhopal plant site.... Dow's sole and unique responsibility is to its shareholders, and Dow CANNOT do anything that goes against its bottom line unless forced to by law." For a while, this as reprinted in something called "Men's News Daily" becomes the top story on news.google.com.

"Whatever be the circumstances under which the news was aired, we will get $12 billion from Dow sooner than later," one Bhopali activist is quoted as saying. But the "false hope" question does come up in some articles. Much as we try to convince ourselves it was worth it, we cannot get rid of the nagging doubt. Did we deeply upset many Bhopalis?

If so, we want to apologize. We were trying to show that another world is possible, and may not have done it so well.... Throughout the day, we are deluged with email, almost all of it positive. Later, the BBC calls again: they want us back at the studio.

Yeah, right! No, really. They want us on for another show, to talk about what has happened. Against our better judgment we go and arrive to find four smiling staffers. "Where are the gendarmes?" Andy asks, and the staffers actually laugh. Another interview on Channel 4, and the day is finally over. Now all we can do is wait to see how it all pans out. Will our fondest hopes be met? Will Dow be forced to concede? Or will the people of Bhopal have to wait twenty more years?

Visit Bhopal.net and help them keep the pressure on Dow.

Source: http://www.theyesmen.org/hijinks/dow/.

13

Do We All Live in Bhopal?

Terribly unique and unsettlingly normal

Accidents happen regularly, though usually without the drama and singularity of this disaster. As the first excerpt in this chapter describes, the worst industrial disaster in American history was also courtesy of the Union Carbide Corporation – the Hawk's Nest Tunnel fiasco of 1930. But ubiquitous as chemical accidents are, Bhopal, and the hauntingly tangible nightmare of the silent, midnight gas, remains an unforgettable template for understanding both the preconditions and the consequences of these accidents. Similarly, the aftermath has echoed universal fears about chemical exposure – about the unknown effects of the chemical environment.

In 1985, George Bradford (a.k.a. David Watson) in an article entitled *We All Live in Bhopal* articulated both the economic conditions that had made Bhopal the site of this unprecedented disaster and the universality of the danger to life and health from chemical contaminants. The potential danger evoked by the phrase was powerful; Bhopal is as far away as the factory down the road, the air you breathe as you fall asleep. The apparent universality of the need for a safe environment allows nearly everyone to identify with the Bhopali survivor.

But the chain of small, disastrous actions and accidents that caused the gas release in Bhopal would not have caused the same reaction in the sister plant in Institute, West Virginia. West Virginia is not Bhopal – the safety systems in the two factories were designed differently, and the US plant had a higher safety rating (see Carbide's internal documents, in "Predictions"). However, just as the Union Carbide factory in Bhopal was built in the older, poorer section of the city – so is its sister plant built in a poor, mostly minority community in West Virginia. We do not all live in Bhopal, and we do not all live in Institute, West Virginia. It remains that lives that were less valued before the disaster are still most vulnerable to industry and to chemistry.

And as the differential in wealth – and hence in vulnerability – between the richest and the poorest worldwide (and within most countries) continues to widen, more and more communities are truly poised on the brink of another Bhopal. As a result, many groups and thinkers have used Bhopal to articulate the human rights to environmental safety, and to comprehend the problems of corporate accountability. Several of these efforts, including the *Charter on Industrial Hazards and Human Rights*, and *Greenpeace's Bhopal Principles on Corporate Accountability*, are excerpted here. The charter (full text available on-line – see source) uses Bhopal to comprehend the rights to a safe environment, rights that have never before been so necessary. Greenpeace has taken Bhopal up fairly recently as well, focusing their activism primary on the contamination issue (see the chapter, "Contamination"). However, they are also using Bhopal as a prism for understanding the problem of making all corporations accountable.

The intent of this chapter is to understand the effect that Bhopal – in both its acute and sustained phases – has had on concepts of rights, crimes and justice. It continues to be a site of suffering, frustration and innovation, but it must simultaneously continue to inform a wider understanding of the dangers and frontiers of environmental contamination, corporate accountability, and disaster response. This chapter – and this anthology – concludes with Amnesty International's twentieth anniversary report on Bhopal. The very fact of the Amnesty report – its first on Bhopal, and its first major report on corporate abuse of human rights – underlines the current urgency in Bhopal. A just resolution to this issue in particular remains pressing – for the survivors of gas and contamination, and for a safer, more just world.

Hawk's Nest: Carbide's American disaster (1986)

Union Carbide also caused the worst industrial disaster in the United States. Excerpted here is a brief account of the Hawk's Nest Tunnel in West Virginia in 1930 as told by Martin Cherniack.

A half century ago in a remote place, more than seven hundred persons may have died to build the Hawk's Nest Tunnel. Those who succumbed were not overwhelmed by some medieval plague: the mysteries of rock dust were sustained by public and industrial policy, not by technical ignorance. Hawk's Nest was born in the transformations of a new industrial age, so judgment of the events there must be based on unrelated criteria: the fate of the men, the companies that created the tunnel, and the future recognition of industrial disease. There is a symmetry in this study without names. Its subject, industry of scale, is also an anonymous process, consisting of great numbers of men and great numbers of replaceable mechanical parts, sometimes treated without distinction. As such, analogies may summon the language of war rather than that of economy. As when hearing the toll of soldiers dead in war, it is natural to ask about the men of Gauley Bridge, "What did they win?"

For themselves the workers won nothing at all except a brief respite from the poverty that afflicted millions of others during the Great Depression. There was no glory, except for Union Carbide, which boasted for years that the Hawk's Nest hydroelectric plant could illuminate the entire city of Charleston. There was certainly no sympathy or gratitude. The trials, the hearings, and the sometimes unfavorable national attention they brought were as unwelcome to the citizens of Gauley Bridge as they were to Union Carbide and its contractor. The migrant workers, looked upon from the first with suspicion and contempt, became the "undesirables, mainly Negroes" who were the real source of the troubles. And, thanks to the legislators and courts of West Virginia, there was no recompense for misery or even death, but for the meager settlements paid by Union Carbide's contractor to a few out of the thousands of workers. For the rest – or so it was widely reported and is still remembered – when they could no longer work they were evicted from the camps and driven away by force. On the first threat of suits, it was said, at least one of the camps was burned to prevent its use by litigants who waited in the area in the hope of vindication in the courts. For the smaller number of men whose homes were nearby, the prospects of local work were diminished by the reluctance of coal mine operators to hire men who had been on the tunnel and who were allegedly more inclined to illness from dusty work.

If even the surviving workers gained little but misery by their toil in the tunnel, what of their employers? Rinehart and Dennis could boast of having built, with extraordinary speed and efficiency, a hydroelectric complex that was and remains an engineering marvel. Yet deaths, court costs, and settlements took a heavy toll on a company whose principal resource was an army of cheap migrant laborers. Rinehart and Dennis had established an enviable reputation by its construction of many tunnels, dams, and power stations before it broke ground on the tunnel that would be its greatest engineering challenge and success. Yet the company would never again compete for a major project. Within five years its assets were largely liquidated, its successors being purely local firms in Charlottesville, Virginia.

What, then, of the company that owned it all? The Union Carbide and Carbon Corporation had conceived the vast project and planned every detail of the dam, the tunnel, the power station, and the plant at Alloy to which they would supply a never-ending flow of electrical power. It had hired the contractor, dictated schedules and provisions for the workers' health and protection, and closely supervised every aspect of the operation. It had also taken care to separate itself and its ephemeral representative, the new Kanawha Power Company, from legal responsibility for the fate of the workers. What did it have to show for its vision and enterprise?

For a time its grandiose dreams were fully realized, and perhaps surpassed. Before the First World War the inventive plans of Major Moreland, founder of the Electro-Metallurgical Company, had overreached themselves. The modest diversion of the Kanawha River at Glen Ferris for the electrically powered molding of high-temperature alloys was a commercial failure. There were no markets for the company's aluminum, though the discarded product would find a use as grave markers above the dead of Fayette County. But Moreland's dreams had lived on in the ambitions of the successor company, challenging mountains and the natural limits of rivers and muscle. Union Carbide had predicted that the Hawk's Nest hydroelectric project would require seventeen years of continuous operation to redeem its initial investment. The costs of construction were paid for in half the time. A new war had meanwhile created a demand for the metals that Union Carbide produced at Alloy, and the muck and the rock extracted from Gauley Mountain were fashioned into the steels and metals of triumphant enterprise. In just forty years the crude, semi-industrial operation at Glen Ferris had been transformed, by the Alloy plant and the Hawk's Nest Tunnel, into the center of the American ferrosilicon industry, employing forty-four hundred persons and powering twenty great furnaces.

From the standpoint of the Union Carbide Company, the Hawk's Nest Tunnel had more than justified itself. There was a unique mismatch between a primitive, poorly paid, and unprotected labor force and an industry ahead of its time, able to utilize the most modern equipment and techniques. The proof of industrial precocity was the Alloy plant itself, which would remain competitive for half a century. It is also in this mismatch between a highly mechanized technology and unskilled physical labor that much of the tragedy at Gauley Bridge can be understood.

Source: Martin Cherniack, *The Hawk's Nest Incident: America's Worst Industrial Disaster*, New Haven: Yale University Press, 1986, p. 106–108.

We all live in Bhopal (1985)

This essay, published by George Bradford (a.k.a. David Watson) in The Fifth Estate, *the anarchist quarterly, was an early articulation of the global significance of the world's worst industrial disaster.*

The cinders of the funeral pyres at Bhopal are still warm and the mass graves still fresh, but the media prostitutes of the corporations have already begun their homilies in defense of industrialism and its uncounted horrors. Some 3,000 people were slaughtered in the wake of the deadly gas cloud, and 20,000 will remain permanently disabled. The poison gas left a 25 square mile swathe of dead and dying people and animals as it drifted southeast away from the Union Carbide factory. "We thought it was the plague," said one victim. Indeed it was: a chemical plague, an industrial plague. Ashes, ashes, all fall down!

A terrible, unfortunate, "accident", we are reassured by the propaganda apparatus for Progress, for History, for "Our Modern Way of Life". A price, of course, has to be paid – since the risks are necessary to ensure a higher Standard of Living, a Better Way of Life.

The industrialization of the Third World is a story familiar to anyone who takes even a glance at what is occurring. The colonial countries are nothing but a dumping ground and pool of cheap labor for capitalist corporations. Obsolete technology is shipped there along with the production of chemicals, medicines and other products banned in the developed world. Labor is cheap, there are few if any safety standards, costs are cut. But the formula of cost benefit still stands: the costs are simply borne by others, the victims of Union Carbide, Dow, and Standard Oil.

Chemicals found to be dangerous and banned in Europe and the U.S. are produced instead overseas – DDT is a well known example of an enormous number of such products, such as the unregistered pesticide Leptophos exported by the Velsicol Corporation to Egypt which killed and injured many Egyptian farmers in the mid 1970s. Other products are simply dumped on Third World markets, like the mercury-tainted wheat, which led to the death of as many as 5,000 Iraqis in 1972, wheat which had been exported from the U.S. Another example was the wanton contamination of Nicaragua's lake Managua by a chlorine caustic soda factory owned by Pennwalt Corporation and other investors, which caused a major outbreak of mercury poisoning in a primary source of fish for the people living in Managua.

Union Carbide's plant at Bhopal did not even meet U.S. safety standards according to its own safety inspector, but a U.N. expert on international corporate behavior told the *New York Times*, "A whole list of factors is not in place to insure adequate industrial safety" throughout the Third World. "Carbide is not very different from any other chemical company in this regard." According to the *Times*, "In a Union Carbide battery plant in Jakarta, Indonesia, more than half of the workers had kidney damage from mercury exposure. In an asbestos cement factory owned by the Manville Corporation 200 miles west of Bhopal, workers in 1981 were routinely covered with asbestos dust, a practice that would never be tolerated here." (12/9/84)

Some 22,500 people are killed every year by exposure to insecticides – a much higher percentage of them in the Third World than use of such chemicals would suggest. Many experts decry the lack of an "industrial culture" in the "underdeveloped" countries as a major cause of accidents and contamination. But where an "industrial culture" thrives, is the situation really better?

In the advanced industrial nations an "industrial culture" (and little other) exists. Have such disasters been avoided as the claims of these experts would have us believe?

Another event of such mammoth proportions as those of Bhopal would suggest otherwise – in this case, industrial pollution killed some 4,000 people in a large population centre. That was in London, in 1952, when several days of "normal" pollution accumulated in stagnant air to kill and permanently injure thousands of Britons.

Then there are the disasters closer to home or to memory, for example, the Love Canal (still leaking into the Great lakes water systems), or the massive dioxin contaminations at Seveso, Italy and Times Creek, Missouri, where thousands of residents had to be permanently evacuated. And there is the Berlin and Farro dump at Swart Creek, Michigan, where C-56 (a pesticide by-product of Love Canal fame), hydrochloric acid and cyanide from Flint auto plants had accumulated. "They think we're not scientists and not even educated," said one enraged resident, "but anyone who's been to high school knows that cyanide and hydrochloric acid is what they mixed to kill people in the concentration camps."

A powerful image: industrial civilization as one vast, stinking extermination camp. We all live in Bhopal, some closer to the gas chambers and to the mass graves, but all of us close enough to be victims. And Union Carbide is obviously not a fluke – the poisons are vented in the air and water, dumped in rivers, ponds and streams, fed to the animals to go to market (mad cows in a mad world), sprayed on lawns and roadways, sprayed on food crops, every day, everywhere. The result may not be as dramatic as Bhopal (which then almost comes to serve as a diversion, a deterrence machine to take our minds off the pervasive reality which Bhopal truly represents), but it is deadly. When ABC News asked University of Chicago professor of public health and the author of "The Politics of Cancer", Jason Epstein, if he thought a Bhopal style disaster could occur in the U.S., he replied: "I think what we're seeing in America is far more slow – not such large accidental leakages with the result that you have an excess of cancers or reproductive abnormalities."

In fact, birth defects have doubled in the last 25 years. And cancer is on the rise. In an interview with the *Guardian*, Hunter College professor David Kotelchuck described the "cancer atlas" maps published in 1975 by the Department of Health, Education & Welfare. "Show me a red spot on these maps and I'll show you an industrial center of the U.S.," he said. "There aren't any place names on the maps but you can easily pick out concentrations of industry. See, it's not Pennsylvania that's red, its just Philadelphia, Erie and Pittsburg. Look at West Virginia here, there's only two red spots, the Kanawha Valley, where there are nine chemical plants including Union Carbide's and this industrialized stretch of the Ohio River. It's the same story wherever you look."

There are 50,000 toxic waste dumps in the U.S. The EPA admits that 90% of the 90 billion pounds of toxic waste produced annually by U.S. industry (70 % of it by chemical companies) is disposed of "improperly" (although we wonder what they would consider "proper" disposal). These deadly products of industrial civilization – arsenic, mercury, dioxin, cyanide, and many others – are simply dumped, "legally" and "illegally" wherever is convenient to industry. Some 66,000 different compounds are used in Industry. Nearly a billion tons of pesticides and herbicides comprising 225 different chemicals were produced in the U.S. last year (1984), and an additional 79 million pounds were imported. Some 2% of chemical compounds have been tested for side effects. There are 15,000 chemical plants in the U.S., daily manufacturing mass death.

All of the dumped chemicals are leaching into our water. Some three to four thousand wells, depending on which government agency you ask, are contaminated or closed in the U.S. In Michigan alone, 24 municipal water systems have been contaminated, and a thousand sites have suffered major contamination. According to the *Detroit Free Press*, "The final toll could be as many as 10,000 sites" in Michigan's "water wonderland" alone (14/4/84).

The cover-ups go unabated here as in the Third world. One example is that of dioxin; during the proceedings around the Agent Orange investigations, it came out that Dow Chemical had lied all along about the effects of dioxin. Despite research findings that dioxin is "exceptionally toxic" with "a tremendous potential for producing Chloracne and systematic injury," Dow's top toxicologist, V. K. Rowe, wrote in 1965, "We are not in any way attempting to hide our problems under a heap of sand. But we certainly do not want to have any situations arise which will cause the regulatory agencies to become restrictive."

The corporate vampires are guilty of greed, plunder, murder, slavery, extermination and devastation. We should avoid any pangs of sentimentalism when the time comes for them to pay for their crimes against humanity and the natural world. But we will have to go beyond them, to ourselves: subsistence, and with it culture, has been destroyed. We have to find our way back to the village, out of industrial civilization, out of this exterminist system.

The Union Carbides, the Warren Andersons, the "optimistic experts" and the lying propagandists all must go, but with them must go the pesticides, the herbicides, the chemical factories and the chemical way of life which is nothing but death. Because this is Bhopal, and it's all we've got. This "once nice place" can't be simply buried for us to move on to another pristine beginning. We must find our way back to the village, or as the North American natives said, "back to the blanket," and we must do this not by trying to save an industrial civilization, which is doomed, but in that renewal of life that must take place in its ruin. By throwing off this Modern Way of Life, we won't be "giving things up" or sacrificing, but throwing off a terrible burden. Let us do so soon before we are crushed.

Source: George Bradford, "We All Live in Bhopal," *Fifth Estate*, Winter 1985. Also reprinted in *Questioning Technology: a critical anthology* (London: Freedom Press, 1988).

Charter on Industrial Hazards and Human Rights (1996)

One of the most important efforts to codify and apply the lessons of the Bhopal disaster was this charter on industrial hazards and human rights. Excerpted here are the introduction and first set of articles.

The Charter on Industrial Hazards and Human Rights was drafted in the spirit of learning from the past so that a more humane future may be possible. It is not an official document, but a people's statement. Unlike most human rights documents its content was not determined by diplomatic compromise. Rather, its substance, and hence its authority, derive directly from the collective experience of those who have been forced to live with the consequences of industrial hazards.

Nearly five years in the drafting, the Charter is based on a series of public hearings on the topic of industrial hazards and human rights. The hearings were held in New Haven (USA, 1991), Bangkok (Thailand, 1991), Bhopal (India, 1992), and London (UK, 1994) under the auspices of the Permanent People's Tribunal (PPT).

Articles 1–6 of the International Charter on Industrial Hazards and Human Rights

PART I: RIGHTS OF GENERAL APPLICATION

Article 1 *Non-discrimination*

1. Everyone is entitled to all the rights and freedoms set forth in this Charter, without distinction of any kind, such as race, color, sex, language, religion, nationality, political opinion or affiliation, ethnic or social origin, disability, age, property, sexual orientation, birth, income, caste or any other status.

2. On account of the particular discrimination faced by women, both as waged and unwaged workers, attention should be given to the specific application of the rights stated below where women may be affected.

3. On account of their vulnerability and exploitation in the labor market, special protection should be accorded to children exposed to industrial hazards.

4. On account of the connection between low wages and hazardous working environments and the disproportionate impact of industrial hazards on racial and ethnic minorities, special protection should be afforded low income groups and racial minorities.

Article 2 *Relation to other rights*

The rights in this charter and other human rights, including civil, political, economic, social and cultural rights, are universal, interdependent and indivisible. In particular, freedom from hazards, including the right to refuse hazardous employment and the right to organize against hazards, depends upon the full implementation of social and economic rights, including the rights to education, health and an adequate standard of living.

Article 3 *Right to accountability*

All persons have the right to hold accountable any individual, company or government agency for actions resulting in industrial hazards. In particular, parent companies, including transnational corporations, shall be liable for the actions of their subsidiaries.

Article 4 *Right to organise*

1. All community members and workers have the right to organise with other local communities and workers for the purpose of seeking to ensure a working environment free from hazard.

2. In particular, the right to organise includes:

(a) the freedoms of expression, association and peaceful assembly;
(b) the right to form local, national and international organizations;
(c) the rights to campaign, lobby, educate and exchange information;
(d) the right to form trade unions;
(e) the right to strike or take other forms of industrial action.

Article 5 *Right to appropriate health care*

1. All persons have the right to appropriate health care.

2. In particular, the right to appropriate health care includes:

(a) the right of individuals and groups to participate in the planning and implementation of health care;
(b) the right of equal access of individuals and families to health care the community can afford;
(c) the right to relevant health care services, including where appropriate access to hospitals, neighborhood clinics, specialist clinics, as well as the services of general practitioners, other medical professionals and health care workers drawn from the affected community;
(d) the right to independent information on the relevance and reliability of health care services and treatments including allopathic, ho-

meopathic, nutritional, physiotherapeutic, psychotherapeutic, indigenous and other approaches;

(f) the right to health care systems which recognize and take account of the different ways in which hazards affect women, men and children;

(g) the right to health education;

(h) the development of national, regional and international networks to facilitate sharing of information and experience.

Article 6 *Right of refusal*

1. All communities have the right to refuse the introduction, expansion or continuation of hazardous activities in their living environment.

2. All workers have the right to refuse to work in a hazardous working environment without fear of retaliatory action by the employer.

3. The right to reject inappropriate legal, medical or scientific advice shall not be infringed.

Source: http://www.pan-uk.org/Internat/indhaz/Charter.pdf, please see the full text, available on-line.

Lots of Chemicals, Little Reaction (2004)

Rick Hind and David Halperin look at how the lessons of Bhopal are being forgotten in the current drive for domestic security in the USA.

Washington – While President Bush continues to make terrorism and domestic security the centerpiece of his campaign, he has made little mention of one of the most urgent threats to our safety: the risk that terrorists could cause thousands, even millions, of deaths by sabotaging one of the 15,000 industrial chemical plants across the United States.

The dangers from chemical plant mishaps are clear. According to data compiled by Greenpeace International, the 1984 accident at an Union Carbide insecticide plant in Bhopal, India, has caused 20,000 deaths and injuries to 200,000 people. A terrorist group could cause even greater harm by entering a plant in the United States and setting off an explosion that produces a deadly gas cloud.

The administration knows the dangers. Soon after the 9/11 attacks, Senator Jon Corzine, Democrat of New Jersey, highlighted the issue with legislation requiring chemical plants to enhance security and use safer chemicals and technologies when feasible. (Such safer substitutes are widely available.)

A study by the Army surgeon general, conducted soon after 9/11, found that up to 2.4 million people could be killed or wounded by a terrorist attack on a single chemical plant. In February 2003, the government's National Infrastructure Protection Center warned that chemical plants in the United States could be Qaeda targets. Investigations by *The Pittsburgh Tribune-Review* and the CBS program "60 Minutes" have highlighted lax or nonexistent security at chemical plants, with gates unlocked or wide open and chemical tanks unguarded.

The Environmental Protection Agency under Christie Whitman did its part to evaluate the threat, identifying 123 chemical facilities where an accident or attack could threaten more than a million people, and 7,605 plants that threatened more than 1,000 people. The agency determined that it could use the Clean Air Act to compel chemical plants to increase security.

Following the Corzine approach, the agency also planned to promote the use of less hazardous chemicals. But the Bush administration overruled the initiative, and in December the president announced that chemical security was now the province of the new Department of Homeland Security, under Secretary Tom Ridge.

As *The Wall Street Journal* disclosed last month, Homeland Security tried to reduce the threat of catastrophic attack with the stroke of a pen. The department announced that the number of plants that threatened more than 1,000 people was actually only 4,391, and the number that endangered more than a million people was not 123 but two.

Mr. Ridge has set in motion plans to install security cameras at chemical plants in seven states – but not in some high-threat states like Florida, Ohio and Minnesota. Although the department visits plants and offers advice, unlike the E.P.A., it doesn't have the power to enforce security measures and relies instead on voluntary efforts by the industry. Without enforceable requirements, chemical firms will remain reluctant to put sufficient safeguards in place, for fear that their competitors will scrimp on security and thus be able to undercut them on price.

Industry groups have lobbied intensely against the Corzine legislation. While reluctant to invest in plant safety, some of these companies and their executives have found the resources to help pay for the Republican campaign.

For the Bush administration, it seems, homeland security is critical except when it conflicts with the wishes of supporters who own chemical plants.

Rick Hind is legislative director of Greenpeace's toxics campaign. David Halperin, a lawyer, has served on the staffs of the National Security Council and the Senate Intelligence Committee.

Source: *New York Times*, op-ed, September 22, 2004.

Greenpeace's Bhopal Principles on Corporate Accountability (2002)

The principles on corporate accountability that Greenpeace presented to the 2002 Earth Summit were named after Bhopal, because internationally Bhopal has come to exemplify corporate crime.

1. Implement Rio Principle 13.

States shall as a matter of priority enter into negotiations for a legal international instrument, and adopt national laws to operationalize and implement Principle 13 of the Rio Declaration, to address liability and compensation for the victims of pollution and other environmental damage.

2. Extend corporate liability

Corporations must be held strictly liable without requirement of fault for any and all damage arising from any of their activities that cause environmental or property damage or personal injury, including site remediation. Parent companies as well as subsidiaries and affiliated local corporations must be held liable for compensation and restitution. Corporations must bear cradle to grave responsibility for manufactured products. States must implement individual liability for directors and officers for actions or omissions of the corporation, including for those of subsidiaries.

3. Ensure corporate liability for damage beyond national jurisdictions

States shall ensure that corporations are liable for injury to persons and damage to property, biological diversity and the environment beyond the limits of national jurisdiction, and to the global commons such as atmosphere and oceans. Liability must include responsibility for environmental cleanup and restoration.

4. Protect Human rights

Economic activity shall not infringe upon basic human and social rights. States have the responsibility to safeguard the basic human and social rights of citizens, in particular the right to life; the right to safe and healthy working conditions; the right to a safe and healthy environment; the right to medical treatment and to compensation for injury and damage; the right to information and the right of access to justice by individuals and by groups promoting these rights. Corporations must respect and uphold these rights. States must ensure effective compliance by all corporations of these rights and provide for legal implementation and enforcement.

5. Provide for public participation and the right to know

States shall require companies routinely to disclose to the public all information concerning releases to the environment from their respective facilities as well as product composition. Commercial confidentiality must not outweigh the interest of the public to know the dangers and liabilities associated with corporate outputs, whether in the form of pollution by-products or the product itself. Once a product enters the public domain there should be no restrictions on public access to information relevant to environment and health on the basis of commercial secrecy. Corporate responsibility and accountability shall be promoted through environmental management accounting and environmental reporting which gives a clear, comprehensive and public report of environmental and social impacts of corporate activities.

6. Adhere to the highest standards

States shall ensure that corporations adhere to the highest standards for protecting basic human and social rights including health and the environment. Consistent with Rio Declaration Principle 14, States shall not permit multinational corporations to deliberately apply lower standards of operation and safety in places where health and environmental protection regimes, or their implementation, are weaker.

7. Avoid excessive corporate influence over governance

States shall cooperate to combat bribery in all its forms, promote transparent political financing mechanisms and eliminate corporate influence on public policy through election campaign contributions, and/or non-transparent corporate-led lobby practices.

8. Protect Food Sovereignty over Corporations

States shall ensure that individual States and their people maintain sovereignty over their own food supply, including through laws and measures to prevent genetic pollution of agricultural biological genetically engineered organisms and to prevent the patenting of genetic resources by corporations.

9. Implement the precautionary principle and require environmental impact assessments

States shall fully implement the Precautionary Principle in national and international law. Accordingly, States shall require corporations to take preventative action before environmental damage or heath effects are incurred, when there is a threat of serious or irreversible harm to the environment or health from an activity, a practice or a product. Governments shall require companies to undertake environmental impact assessments with public participation for activities that may cause significant adverse environmental impacts.

10. Promote clean and sustainable development

States shall promote clean and sustainable development, and shall establish national legislation to phase out the use, discharge and emission of hazardous substances and greenhouse gases, and other sources of pollution, to use their resources in a sustainable manner, and to conserve their biological diversity.

Source: Presented by Greenpeace at the Johannesburg Earth Summit – 2002, www.greenpeace.com.

Redefining the public debate on chemical security (2004)

The U.S. Chemical Safety and Hazard Investigation Board has been analyzing serious accidents in U.S. chemical plants and other industrial facilities. Here are some disturbing findings from CSB investigations.

Workers Be Damned: The Bush and ACC team didn't stop at the EPA or the U.S. Congress.

After he took office, Mr. Bush resolved a long-standing issue for the ACC. In an October 1997 presentation to OSHA, four industry trade organizations, including the Chemical Manufacturers Association (CMA now renamed ACC) and the Synthetic Organic Chemical Manufacturers Association (SOCMA) jointly urged OSHA "not to expand the Process Safety Management (PSM) rule in an attempt to address the very difficult issue of reactive hazards." [18] The PSM rule was aimed at enhancing occupational safety.

Finally, in 2002 with help from the White House, the industry got its wish. Ironically, on December 3, 2002 – the 18th anniversary of the Union Carbide disaster in Bhopal – a note in the Federal Register revealed the chemicals initiative was being dropped because of "resource constraints and other problems." By February 2003, OSHA's budget was slashed by $7.9 million and its workforce reduced by 83 jobs, according to the Center for Public Integrity's report.

Later, the public interest group OMB Watch, found that the chemicals initiative was part of a non-public industry "hit list" of 57 regulations comprising items that the industry considered too burdensome.

Job Security = Chemical Security

Responding to a 1994 survey, the *Industrial Safety and Hygiene News* magazine found three-quarters of the respondents – mostly industry personnel – said business competition and downsizing is forcing them to cut safety spending. Nearly 80 percent believed accidents are more likely as employees work longer hours, handle new assignments and fear for their jobs.[27]

A March 2003 GAO report "Voluntary Initiatives are Under Way at Chemical Facilities but the Extent of Security Preparedness is Unknown" cites industry officials to conclude that "...chemical companies must weigh the cost of implementing countermeasures against the perceived reduction in risk." The Conference Board reports "The perceived need to upgrade corporate security has clashed with the perceived need to control expenses until the economy recovers."[28] Workers are often the first victims when safety is compromised to cut costs. The most significant among the unlearned lessons from the Union Carbide disaster in Bhopal is the important role of workers. It is undisputed that Union Carbide systematically ignored workers' warnings about plant safety and even fired workers that raised such issues. During the cost-cutting drive that preceded the disaster, Union Carbide laid off safety personnel and other workers, reduced safety training and cut back expenditure on safety systems. "Maintenance personnel [at the fateful methyl isocyanate unit] were cut from six to two. In 1983, the fire and rescue squad was filled with unqualified persons and later changed to an emergency squad. One month prior to the disaster, the vital post of maintenance supervisor was eliminated from the night shift."[29] The rest of what happened at midnight on December 2-3, 1984, is history. As the CSB's Poje observes, "The analogy of Bhopal still resonates here because of the endangerment due to downsizing."[30] Drawing upon limited data relating to the period 1992-2001, when 91 deaths of contract employees were known to have resulted from explosions or asphyxiation in industrial plants, the Center to Protect Workers' Rights found many instances of injury/death were caused by the lack of workers' knowledge of the presence of explosive material, or the lack of adequate precaution.[31] The report recommends "the use of outside contractors working in industrial plants should be reviewed to determine unique safety risks and needs for this group."

Notes:

18. "OSHA PSM Revisions: Discussions with OSHA." American Petroleum Institute, Chemical Manufacturers Association, Organization Resources Counselors, Synthetic Organic Chemical Manufacturers Association. October 1997.
27. Industrial Safety and Hygiene News. May 1994. pp.3132. Cited in "Accidents Waiting to Happen: Hazardous Chemicals in the U.S. Fifteen Years After

Bhopal." Jeremiah Baumann (U.S. PIRG Education Fund); Paul Orum (Working Group on Community Right to Know); Richard Puchalasky (Grassroots Connection). December 1999. 28. "Corporate Security Management: Organization and Spending Since 9/11." Conference Board. 2004.

29. "Bhopal: The Inside Story. Carbide Workers Speak Out on the World's Worst Industrial Disaster." T.R. Chouhan and Others. The Apex Press, NY; Other India Press. Goa. 1994.

30. Phone interview with Gerald Poje, US Chemical Safety Board. 3 June, 2004.

31. "Explosions and Asphyxiation Deaths Among Contract Employees in Industrial Plants." Michael McCann. Center to Protect Workers' Rights. December 2003. Environmental Health Perspectives 110:721-728 (2002).

Source: No More Bhopals Alliance, "Chemical Industry v. Public Interest: Redefining the Public Debate on Chemical Security" June 2004, http://www.come-clean.org/pdfs/security.pdf

Amnesty International's report on Bhopal (2004)

On the twentieth anniversary, Amnesty International added its voice to the chorus of survivors and organizations, including Greenpeace, and followed shortly after by chapters of Friends of the Earth, demanding justice and rehabilitation in Bhopal. This is their conclusion.

In 1989, cutting short ongoing legal proceedings, the Indian Supreme Court announced a court-endorsed final settlement between the corporation and the government of India without consulting the victims. It said that providing relief to victims took precedence over settling questions of law and liability. In response to a modest financial payment to victims, the settlement bestowed sweeping civil and criminal immunity on UCC, trading off its legal liability while excluding the victims of the disaster from shaping the end of the case. The payment of compensation did not, however, begin until 1992 and involved numerous problems including payment of inadequate sums, delayed payments and arbitrary rejection of claims.

In 1994, all government research on the medical effects of the Bhopal disaster was discontinued without explanation. The full results of the research carried out have yet to be published.

Government efforts to provide rehabilitation have proved ineffective. The poor quality of the health care system has meant that most survivors have had to spend most of their compensation money on medical treatment. Economic rehabilitation measures have failed to prevent the impoverishment of already economically vulnerable survivors.

The report concludes that there is no substitute for taking steps to regulate the activities of transnational corporations in both host and home countries. Laws in host countries must be developed and enforced to allow national governments and local communities to control the activities of transnational companies operating in their territory. Transnational corporations should avoid double standards in safety and adopt the best practices in all aspects of safety in all their operations.

The Bhopal disaster and its aftermath demonstrate clearly the need for an international human rights framework that can be applied to companies directly, that could act as a catalyst for national legal reform, and could serve as a benchmark for national law and regulations. Ensuring public participation and transparency in decisions relating to the location, operational safety and waste disposal of industries using hazardous materials and technology is an essential step to heighten risk awareness and responsible behavior as well as to ensure better preparedness to prevent and deal with disasters like Bhopal.

The international community must ensure that victims of human rights violations have effective access to justice and effective redress for the harm suffered, without discrimination, and regardless of whether those responsible for the violations are governments or corporations.

Source: *Clouds of Injustice: Bhopal 20 years on*, Amnesty International, December 2004.

Conclusion

HOW HAS THIS particular tragedy managed to slip through the cracks of the systems of responsibility designed for abuses of any of these categories – legal, environmental, medical, corporate and human? The categories of personal injury, environmental contamination and negligence, culpable homicide, and human rights become inadequate in the face of the Bhopal disaster. Abandoned in a gray area of national and international law that applies (or doesn't) to multinational corporate actors, Bhopal and its aftermath continue to demand a radical revision of international justice and corporate accountability.

When gas leaked from the Union Carbide plant in Bhopal in 1984, it radically changed the public understanding of the chemical industry. More significant however, is what, in the course of the 20 years since, it has *not* changed: none of the trials for personal criminal responsibility has been resolved; the issue of water contamination near the site is still unresolved; the hospitals funded by UC do not comprehensively address the health concerns of the survivors and their children, and there has still been no admission of responsibility – by anyone. In short, there has been no shift in the balance of power between corporations and those who survive their mistakes.

What does it mean to "remember" something that has not ended? In the Union Carbide corporation's film entitled "Unraveling the tragedy," a wistful lawyer at a conference table comments that, had not the government interfered in Carbide's investigation, "the way that people remember Bhopal today might be different." Indeed, and had not Union Carbide spent millions of dollars on public relations, misinformation and evasion, perhaps the lives of the affected people would be different – and perhaps more of them would be alive today to tell it.

To "remember Bhopal" today means not just collecting and understanding information about the disaster and its aftermath, but also acting on it and using it in creative ways. The way that we remember Bhopal *should* be different. Through this work of memory and advocacy, we, in solidarity with those struggling for health, survival and justice in Bhopal, are working for a future memory of Bhopal that is *not* part of a continuing tragedy.

Resources

Select Bibliography

*Amnesty International (2004). *Clouds of Injustice*, New York, NY.

Baxi, U. (1986). *Inconvenient Forum and Convenient Catastrophe: The Bhopal case*. Bombay, N.M. Tripathi.

Baxi, U. a. Dhanda, A. (1994). *Valiant Victims and Lethal Litigation: The Bhopal case*. Bombay, N.M. Tripathi.

Baxi, U. a. P., Thomas (1986). *Mass Disasters and Multinational Liability: The Bhopal case*. Bombay, N.M. Tripathi.

Bogard, William (1989). *The Bhopal Tragedy: Language, logic, and the politics of a hazard*, Westview Press, Boulder, CO.

Cassells, J. (1993). *The Uncertain Promise of Law: Lessons from Bhopal*, University of Toronto Press.

Cherniak, Martin. (1986). *The Hawk's Nest Incident: American's worst industrial disaster*, Yale University Press, New Haven, CT.

*Chouhan, T.R. and others (2004). *Bhopal: The inside story*, The Apex Press, New York, NY, and Other India Press, Goa, India.

De Grazia, Alfred (1985). *A Cloud Over Bhopal: Causes, consequence, and constructive solutions*, Kalos Foundation, Bombay, India.

*Dembo, David, et al. (1990). *Abuse of Power: Social performance of multinational corporations: The case of Union Carbide*. New Horizons Press, NY.

*Doyle, Jack (2004). *Tresspass Against Us: Dow Chemical & the toxic century*, Common Courage Press, Monroe, ME.

Diamond, Arthur (1990). *The Bhopal Chemical Leak*, World Disaster Series. Lucent Books.

*Eckerman, Ingrid (2004). *The Bhopal Saga : Causes and consequences of the world's largest industrial disaster.* Andhra Pradesh, India, Universities Press.

Everest, L. (1986). *Behind the Poison Cloud: Union Carbide's Bhopal massacre*, Banner Press. http://www.larryeverest.com/book-poison-cloud.htm.

Fortun, K. (2001). *Advocacy after Bhopal: Environmentalism, disaster, new global orders.* Chicago, The University of Chicago Press.

*Hanna, Bridget, Ward Morehouse, and Satinath Sarangi (2005) *The Bhopal*

Reader: Twenty years of the world's worst industrial disaster, The Apex Press, NY and Other India Press, Mapusa, Goa, India.

Jasanoff, S. (1994). *Learning from Disaster: Risk management after Bhopal* (Law in Social Context), University of Pennsylvania Press.

*Lapierre, D. and Javier Moro (2002). *Five Past Midnight in Bhopal.* New York, Warner Books, Inc.,

Larabee, A. (2000). *The Bhopal Effect. Decade of Disaster.* Chicago, University of Illinois Press.

Mooney, B. J. (2002). *The Bhopal Disaster: Discourse and narrative in the ethnography of an event.* Anthropology. Ann Arbor: 292. http://www.il.proquest.com

*Morehouse, Ward and Arun Subramaniam (1986). *The Bhopal Tragedy : What really happened, and what it means for American workers and communities at risk.* New York, Alfred A. Knopf. http://www.cipa-apex.org

*Morehouse, Ward and Chandana Mathur, (2005) *People Against Power: Essays provoked by Bhopal's twentieth anniversary*, Great Barrington, MA: North River Press.

*Mukherjee, S. (2002). *The Bhopal Gas Tragedy, a book for the young reader.* Chennai, Tulika Publishers, http://www.tulikabooks.com/infocus2.htm.

Rai, Raghu. (2002). *Exposure (Photography).* Mumbai, Greenpeace. http://www.aspp.com/gallery/archive20_rai/index.lasso.

Rajan, S. R. (1999). *Bhopal: Vulnerability, routinization and the chronic disaster.* The Angry Earth: Disaster in Anthropological Perspective. A. O.-S. a. S. M. Hoffman. New York and London, Routledge.

Riddle, John (2002). *Bhopal.* Chelsea House.

Sahabat Alam Malaysia (1985) *The Bhopal Tragedy: One Year After*, Appen, Penang, Malaysia.

Shrivastava, Paul (1987). *Bhopal: Anatomy of a Crisis*, Ballinger, Cambridge, MA.

Vivek, P. S. (1990). *The Struggle of Man Against Power: Revelation of 1984 Bhopal tragedy*, Himalaya Pub. House; 1st ed. edition.

*Weir, David (1987). *The Bhopal Syndrome: Pesticides, environment and health.* San Francisco, Center for Investigative Reporting.

Note: Titles marked with an asterisk (*) are from the Bhopal Twentieth Anniversary Library and are available online at www.cipa-apex.org or from The Apex Press, Suite 3C, 777 UN Plaza, New York, NY 10017; 800-316-APEX (2739).

Films*

- 2004 *Litigating Disaster* – by Ilan Ziv

  How is it possible that nearly two decades after an event of such magnitude there is no legal closure? Constructed as attorney Rajan Sharma's case as presented to fictitious jurors, *Litigating Disaster* takes the viewers on a riveting cinematic investigation; presenting the compelling evidence assembled against Union Carbide including unique, never before seen documents unearthed through prolonged legal struggles, exclusive interviews with Union Carbide former officers, powerful archival material, and scenes filmed in and out of the abandoned plant. (52 minutes)

  www.tamouzmedia.com

- 2004 *Bhopal: The Search for Justice* – by Peter Raymont and Lindalee Tracey

  This film straddles the intersection between science, politics and human rights. Exploring charges of corruption, graft and greed, the film follows Rajkumar Keswani, the local journalist whose prediction of the Union Carbide disaster proved prophetic. Set against the rich visual tapestry of India, Keswani travels through the Indian bustees where the poorest victims live, and to the offices of frustrated doctors and scientists. Finally he makes his first trip to North America in search of answers. 52 minutes.

  Website: http://www.onf.ca/bhopal/.

- 2004 *One Night in Bhopal* – by Steven Condie and the BBC

  This film reveals how and why an American-owned chemical factory that was meant to bring prosperity to the people of an Indian city, instead brought death and destruction. By mixing drama, documentary, graphics and archive material, the testimony of key witnesses, this documentary by the BBC reveals the events of the Bhopal disaster and how Union Carbide responded. It also gives a picture of the survivors today and their continuing health effects. 60 minutes.

  Online at http://www.bbc.co.uk/newsa/n5ctrl/progs/04/one_offs bhopal01dec.ram

- 2004 *Twenty Years Without Justice* – by Sanford Lewis

  A campaign video that explores the consequences of the disaster, its causes, and the 20-year campaign which seeks justice for those who survived it. Includes interviews with gas survivors, the lawyer behind their historic lawsuit against Union Carbide, and a former engineer at the now-abandoned factory site. 17 minutes. Online at *http://bhopal.strategicvideo.net.*

- 2004 *The Goldman Awards*

  A documentary narrated by Robert Redford which briefly recounts the events of "that night" and the stories of Rashida Bee and Champa Devi Shukla, two gas-affected women and leaders of the international campaign for justice who were jointly awarded the prestigious Goldman Environmental Prize in April. 5 minutes.

Online at www.goldmanprize.org/press/library.html.

- 2003 *Dateline: "Hunting Warren Anderson"*

  For almost 20 years, the survivors of the disaster have sought punishment for those they hold responsible, and they start at the top. They're targeting the American chief executive of the company – the now-retired 82-year-old Warren Anderson. So where does the buck stop when it comes to culpability for the world's worst industrial tragedy? Amos Cohen reports. A June 18, 2003 episode of the Australian Broadcast Company's news magazine. 35 minutes. Online at http://203.15.102.143:8080/ramgen/media/dl_180603a.rm.

- 2002 *MITGIFT – Aus Katastrophen lernen; Seveso, Bhopal & Co.*, – by Thomas Weidenbach and Meike Hemschemeier

  This piece, screened on WDR television, compares the lessons of chemical danger and disaster in Seveso, Italy, and in Bhopal India. 45 minutes, German, contact: Weidenbach@tweidenbach.de.

- 2001 *Cloud Over Bhopal* – by Gerardo Olivares and Larry Levene

  A short documentary based on Dominique Lapierre's book *Five Past Midnight in Bhopal.* It was five past twelve on the night of the 2nd to the 3rd of December 1984. A dazzling cloud of toxic gas escapes from an American pesticide factory built in the ancient Indian city of Bhopal. The result: thirty thousand dead and five hundred thousand injured. A true story. Gondwana Films, 53 minutes.

- 1999 *The Heart Becomes Quiet* – by Robin Schlaht and David Christensen

  This film examines the legacy of a tragedy that has brought out the worst in some people, yet has offered others the possibility of redemption. The documentary examines seven individuals whose lives have been changed by the gas tragedy. 88 minutes, color, Hindi and Urdu with English subtitles, Copyright Agitprop Films and Zima Junction Productions. www.maplelake.mb.ca.

- 1999 *Bhopal Express* – by Mahesh Mathai

  Bhopal Express, the only feature film ever made about the Bhopal tragedy, explores the true story of one of the world's largest industrial disasters. Championed by David Lynch and featuring Naseeruddin Shah and Zeenat Aman, the tragedy is revealed through the experiences of newlyweds Verma (Kay Kay), a foreman at the Carbide plant, his wife Tara (Nethra Raghuraman) and their friend Bashir (Shah). 100 minutes, Hindi with English subtitles. www.bhopalexpress.com

- 1999 *India & Free Trade: A Closer Look at Bhopal* – by Pavithra Narayanan

  This film examines the Bhopal disaster through the context of increasing globalization and questions the tremendous influence that corporate giants wield throughout the 'developing' world. The documentary was released on the fifteenth anniversary of the Bhopal disaster and has a running time of 35 minutes.

- 1999 *The Seduction of Dr. Loya* – by Priya Krishnaswamy

Tells the story of the disaster through the eyes of Union Carbide's Indian medical officer, who was called to the scene in the immediate aftermath of the leak. It's an exploration of the complexity of his reaction, and his torn loyalties. Union Carbide had paid for him to go to medical school, and gave him a plush job, but his mother died as a result of the leak, as did thousands of others. 45 minutes.

- 1995 *Bhopal: The Second Tragedy* – by Mark Tully

  This documentary follows Mark Tully, the BBC correspondent who reported on the disaster from Bhopal in its immediate aftermath, as he returns to Bhopal ten years after the disaster. (52 minutes)

- 1994 *Bhopal: Setting the Record Straight* – by G. Switkes

  A conversation with Munoz, former Managing Director of Union Carbide India. – An interview with Josh Karliner of Corpwatch. Munoz discusses his objections to the building of large storage tanks to hold MIC at the Bhopal plant. He talks of the dangers inherent in the continued storage of vast quantities of MIC at the Institute, West Virginia, plant. 25 minutes, English. Available from www.corpwatch.org.

- 1991 *Chemical Valley* – by Mimi Pickering and Anne Lewis Johnson

  Chemical Valley begins with Bhopal and the immediate response in the Kanawha Valley, USA, an area once dubbed by residents, "the chemical capital of the world." The program follows events over the next five years, exploring issues of job blackmail, racism, and citizens' right to know and to act as it documents one community's struggle to make accountable an industry that has all too often forced communities to choose between safety and jobs. 58 minutes.

- 1989 *Unraveling the Tragedy at Bhopal*

  This video explains Union Carbide's version of the events. Available from the Union Carbide Corporation. 17 minutes.

- 1987 *Bhopal – What Next?* – by Ian Graham

  A film made by the International Confederation of Free Trade Unions (ICFTU), which includes footage from the aftermath, interviews with trade union officials who predicted the disaster or analyzed the aftermath, a refutation of the sabotage theory, and survivor interviews. 40 minutes.

- 1986 *Bhopal: Beyond Genocide* – Directed by Tapan Bose

  Focusing on the Bhopal gas tragedy, this film presents a tale of technological development gone awry and was temporarily banned by the Indian Film Censor Board. 80 minutes.

- 1986 *Bhopal: License to Kill* – by Shoba Sadogopan, Reena Mohan, and Ranjan Palit

  An independent Indian documentary that reveals the design and safety failures at the Bhopal plant and explains how they contributed to the disaster. The film also includes many interviews with the victims of the disaster, as well as the medical personnel who treated them on the night of the disaster and afterwards. 55 minutes.

*A videography from the first decade of the Bhopal struggle is available at the Bhopal Library at cipa-apex.org.

Online Resources

Activist Organizations

- International Campaign for Justice in Bhopal
  http://www.bhopal.net
- Sambhavna Clinic and the Bhopal Medical Appeal
  http://www.bhopal.org/
- Students for Bhopal
  http://www.studentsforbhopal.org
- Greenpeace
  http://www.greenpeaceusa.org/bhopal http://www.greenpeace.org/international_en/extra/?campaign_id=3991

Archives and Other Information Sources

- Bhopal Memory Project
  http://www.bard.edu/bhopal
- Bhopal Library
  http://www.cipa-apex.org
- Business and Human Rights Archive
  http://www.business-humanrights.org/
- Chemical Industry Archives (project of the Environmental Working Group)
  http://www.chemicalindustryarchives.org/dirtysecrets/bhopal/index.asp
- Endgame – Information about the Union Carbide Corporation
  http://www.endgame.org/carbide.html
- India Together
  http://indiatogether.org/campaigns/bhopal
- Media Coverage
  http://bhopalinthenews.blogspot.com/
- Medical Research Archive
  http://webdrive.service.emory.edu/users/vdhara/papers.htm
- More Bhopals are waiting to happen
  http://infochangeindia.org/
- Music & Art from and for Bhopal
  http://www.bhopal.fm/
- remember-bhopal listserv
  http://lists.essential.org/mailman/listinfo/remember-bhopal

Resources and Links

- http://homepages.gs.net/~aaswell/bhopal
- The Truth About Dow
  http://www.thetruthaboutdow.org
- Water contamination tour
  http://www.bhopal.net/oldsite/contaminationtour.html

Corporate Sites

- Bhopal (by the Union Carbide Corporation)
  http://www.bhopal.com
- Dow Chemical Company
  http://www.dow.com

Reports and Resolutions

- 2004 Shareholder's resolution http://www.dow.com/financial/2004prox/agenda4.htm
- Amnesty International report http://web.amnesty.org/pages/ec-bhopal-eng
- Fact Finding Mission on Bhopal. http://www.bhopalffm.org/
- International Medical Commission on Bhopal http://www.pan-uk.org/pestnews/Pn34/pn34p16.htm
- Permanent Peoples' Tribunal Charter on Industrial Hazards & Human Rights http://www.globalpolicy.org/socecon/envronmt/charter.htm
- Permanent Peoples' Tribunal on Global Corporations and Human Wrongs http://elj.warwick.ac.uk/global/issue/2001-1/ppt.html

---

Survivor and Activist Organizations in Bhopal

- Bhopal Gas Peedit Mahila Stationery Karamchari Sangh
  Mrs. Rashida Bee & Mrs. Champa Devi Shukla Gas Peedit Mahila Stationery Ikai
  Hastshilp Vikas Nigam Parisar
  Hamidia Road, Bhopal
  Madhya Pradesh, 462000
  Tel: 91-9303132298, 9303132959
- Bhopal Ki Aawaaz
  Mr. Shahid Noor
  House No 69, Risaldar Colony
  Chhola Road, Bhopal
  Tel: 91-9303122784
- Bhopal Gas Peedit Mahila Purush Sangarsh Morcha
  Mr. Syed M. Irfan
  House No 7, Lane No 2, Jogipura, Budhwara
  Bhopal, Madhya Pradesh 462000
  Tel: 91-9329026319
- Bhopal Gas Peedit Mahila Udyog Sangathan
  Mr. Abdul Jabbar
  House No. 51, Rajendra Nagar
  Bhopal 462 010
  Tel. 91 755 5242727, 2748688
- Gas Peedit Nirashrit Morcha, Bhopal
  Mr. Balkrishna Namdeo
  542, Housing Board Colony, Aishbag
  Bhopal 462 010
  Tel. 9826345423
- Bhopal Group for Information and Action
  Mr. Satinath Sarangi, Ms. Rachna Dhingra
  B-1/302, Sheetal Nagar Berasia Road
  Bhopal, MP 462018
  Tel. 91-9826167369
  e-mail: icjb@bhopal.net

Support Organizations

- **Association for India's Development**
  contact: Nishant Jain
  12345 Alameda Trace Circle, #924
  Austin, TX 78727
  512-422-7169
  nishj@umich.edu
- **Bhopal Action Resource Center**
  777 United Nations Plaza, Suite 3C
  New York, NY 10017
  http://www.cipa-apex.org/programs.htm

- **Bhopal Information Network, Japan**
  Tani Yoichi who is president of Japan information network for Bhopal
  627-39 Fukuhama Tsunagi
  Kumamoto
  JAPAN 869-5605
  Tel 81-966-78-2137 Fax 81-966-78-4173
  EZG01444@nifty.ne.jp
- **Calhoun County Resource Watch**
  Diane Wilson
  Box 1001
  Seadrift, Texas 77983
  WilsonAlamobay@aol.com
  (H) 361 785 3907
- **Ecology Center of Michigan**
  Tracey Easthope
  tracey@ecocenter.org
  734-663-2400x109
- **Environmental Health Fund**
  Gary Cohen
  Executive Director
  Environmental Health Fund
  41 Oakview Terrace
  Jamaica Plain, MA 02130
  617-524-6018 (Tel)
  617-524-7021 (fax)
  gcohen@igc.org
- **Greenpeace International**
  Zeina Al-hajj
  zeina.alhajj@int.greenpeace.org
- **Groundwork**, South Africa
  Bobby Peek
  bobby@groundwork.org.za
- **La Campagna Italiana per la Giustizia a Bhopal**
  Vincenzo Mingiardi
  minjar@iol.it
- **Pesticide Action Network**
  Diana Ruiz
  (415) 948-4807 cell
  415 981-6205
  Ext. 321 PANNA office
  dianaruiz@panna.org
- **Pesticide Action Network**
  49 Powell Street
  Suite 500
  San Francisco, CA 94102
- **Students for Bhopal**
  Ryan Bodanyi
  401-829-6192
  rbodanyi@studentsforbhopal.org
  177 Vinton Street
  Providence, RI 02909
- **The Other Media, India**
  Anuradha Saibaba
  campaigns@theothermedia.org
- **UK Campaign for Justice in Bhopal**
  Indra Sinha
  indra@indrasinha.com
  Yes-men's Dow parody
  www.dowethics.com, http://www.bhopaldoesntexist.com
- **Young Volunteers for the Environment**, Togo
  Sena ALOUKA Executive Director
  Young Volunteers for Environment (YVE)
  37, rue 218 akossombo (route du marché d'above)
  Box : 80470 Tel : +228- 220 01 12 - 220 00 34
  Mobile: 9216740 / 9125376
  Fax : +228- 251 47 39 Lome, Togo

For **further information** on organizations that support the Bhopal campaign, please visit http://www.studentsforbhopal